Springer Proceedings in Physics

Volume 150

for Shawn
and Hisako

Friedemann
and think
久子

Friedemann Freund · Stephanie Langhoff

Editors

Universe of Scales: From Nanotechnology to Cosmology

Symposium in Honor of Minoru M. Freund

 Springer

Editors
Friedemann Freund
GeoCosmo Science Center
Los Altos, CA, USA

Stephanie Langhoff
NASA Ames Research Center
Moffett Field, CA, USA

ISSN 0930-8989
Springer Proceedings in Physics
ISBN 978-3-319-02206-2
DOI 10.1007/978-3-319-02207-9
Springer Cham Heidelberg New York Dordrecht London

ISSN 1867-4941 (electronic)

ISBN 978-3-319-02207-9 (eBook)

Library of Congress Control Number: 2014938355

Printed on acid-free paper

Springer is part of Springer Science+Business Media (www.springer.com)

Contents

Part I
Memories of a Great Life

Photo Credit: Tomoko Ishihara

Mino 1962–2012

Reflections

Friedemann Freund and Hisako Matsubara

It is not easy to write about one's own son who has passed away in the prime of his life. Mino was such an extraordinary person that even we as parents have to admit that we did not grasp the scope of his personality and the range of his abilities. We failed to recognize his full potential. Therefore we are now left with a nagging feeling that we should have done more and could have done more.

When Mino came to NASA Ames, he blossomed. After ETH Zürich, UC Berkeley, CalTech–JPL, the Japanese Space Research Institute ISAS, the NASA Goddard Space Flight Center, and the Air Force Research Laboratories—traveling from low-temperature physics to cosmology to nanotechnology—he found at the NASA Ames Research Center an environment, where he could combine his knowledge and understanding of science with his keen sense of what it technologically possible. Supported by many who were drawn in by his contagious optimism and quick mind, Mino embarked on a suite of ambitious projects from basic science to new satellite concepts. He promoted ideas, which others had not yet even started dreaming about, such as fleets of hundreds of nanosatellites, fully capable despite their diminutive size, flying in formation, communicating with each other and the ground, transmitting crucial services to all parts of the world. He crisscrossed the country and spent endless hours in Washington to drum up support for his dreams.

Cosmology filled Mino with awe for the vast expanse of the universe. Nanotechnology gave him a glimpse of the smallest dimensions, where the laws of classical physics no longer apply. But his greatest pleasure derived from his love for Nature, watching the sunset over the Pacific Ocean or hiking—often alone—among the giant redwood trees in Big Basin. It filled him with humility for Nature's seemingly endless life forces.

F. Freund · H. Matsubara
GeoCosmo Science Center, Los Altos, CA, USA

F. Freund, S. Langhoff (eds.), *Universe of Scales: From Nanotechnology to Cosmology*,
Springer Proceedings in Physics 150, DOI 10.1007/978-3-319-02207-9_1,
© Springer International Publishing Switzerland 2014

Mino's Smile

Pete Worden

This book is a compendium of scientific presentations at a Symposium held at the NASA Ames Research Center in honor of Dr. Minoru M. Freund and his work. I can't think of a more appropriate way to remember Mino—as we all called him—than revisiting the ideas that he helped formulate and move forward, many of them truly revolutionary. In his congenial way Mino was quite an activist, a zealot and a visionary.

I met Mino a little less than seven years ago. We had our regular board meeting. Mino, who was new, sat in the corner smiling. This smile came to mean to me everything good about what we do at the NASA Center. Mino had that smile every time I saw him, even when he was diagnosed with this terrible disease. He smiled as he fought for his life. He smiled the last time I saw him.

I like to remember Mino first of all as a colleague. He was one of the most brilliant people I've ever met. He was full of ideas. For him everything was possible. He came to my office with probably an idea a week, and unlike most ideas that I hear, his were good. In fact we had to slow him down and try to keep him from starting all of his ideas. But there was one, to which he dedicated an enormous amount of energy. As early as 2006 Mino came up with new ways to put swarms of nano-satellites into orbit, hundreds of them flying in formation, communicating with each other, doing science and providing links to places around the globe with which we can't communicate today. We first laughed a bit about it and joked, coining the word "Mino-Sats". Mino worked very hard at this idea. He set out to convince others of the feasibility of such an ambitious project. He traveled to Washington to build support. He traveled across the United States and Europe. He brought in a little bit of money, enough to start but not enough to bloom. Today, barely seven years later, Mino's vision of distributed swarms of nano-satellites is being pursued around the world. The Mino-Sats are becoming a reality.

Mino believed in the future—the future that NASA represents. I don't remember how often I discussed with him that sometime in this century humans will leave

P. Worden
Director of the NASA Ames Research Center, Moffett Field, CA, USA

F. Freund, S. Langhoff (eds.), *Universe of Scales: From Nanotechnology to Cosmology*,
Springer Proceedings in Physics 150, DOI 10.1007/978-3-319-02207-9_2,
© Springer International Publishing Switzerland 2014

this planet forever. People will live in other worlds. Mino thought hard and long about how to achieve this goal. Many of the things Mino initiated as the Director of the Center for Nanotechnology were in anticipation of such a time. He had a deep understanding of the technologies that would be needed and he moved forward, ever optimistically. But he also moved forward to make life on Earth better and safer. He brought me Friedemann's ideas about how we could forecast earthquakes days, even weeks before they strike and thereby save lives. Mino also was the first to talk to me about Life Sciences and how important Life Science is for our future—our future in space and our future on earth. To the day that I last saw Mino, he was full of ideas.

But last, and maybe most important, Mino was a friend. Occasionally, when I was a little bit gruff, Mino would come in with a smile. One of my fondest memories is when he said: "You look like you need a hike." Then we'd go out hiking either along the beach or hiking among the Giant Redwoods. Mino radiated an element of peace, the peace of a good friend, a good colleague, somebody who cared about you. Mino never came to my office asking for something for himself. He always asked: "How are you doing?" "How can I help you?" "How could I make you feel better?"

So, on this occasion as we remember Mino, I think of the poem by Hisako, Mino's mother: Mino is with us, and he will be with us as long as we remember him.

Mino, wherever you are: Godspeed to your future and the future that you imagined for us.

Mino in Adjectives

Steve Zornetzer

I want to share with you words that come to my mind as I think of Mino:

Unpretentious
Dreamer
Energetic
Idealistic
Hard-headed
Direct
Curious
Adventurous
Persistent
Entrepreneurial
Courageous
Cultured
Empirical
Generous
Imaginative
Independent
Self-assured
Determined
Brave
Tireless...

That was the Mino I knew.

S. Zornetzer
Associate Director of the NASA Ames Research Center, Moffett Field, CA, USA

F. Freund, S. Langhoff (eds.), *Universe of Scales: From Nanotechnology to Cosmology*,
Springer Proceedings in Physics 150, DOI 10.1007/978-3-319-02207-9_3,
© Springer International Publishing Switzerland 2014

With Mino on the Road

Alan Weston

Mino was a true dreamer and a great friend of mine. I will never forget his cheery disposition, his never-ending smile. When I remember Mino, I am struck by the fact that he is the only person I have ever met who never was upset, no matter what.

Mino was a great team player and team leader. He inspired his colleagues with his energy, enthusiasm and dedication to finding solutions that would benefit humanity. Self motivated but selfless, he went to extraordinary lengths to guide and protect his research team in the turbulent and often unpredictable world of NASA funding for advanced technology.

Mino and I shared a passion for small, even tiny spacecraft that could be built and operated at costs that are so low that they will revolutionize space. Mino's background in nanotechnology led him to question the assumptions of the space community, leading him to conceive of architectures that will forever change humanity, for example by orbiting inter linked nano spacecraft that can provide internet access to the poorest and most remote regions of the planet.

Together we pursued this dream, and I have fond memories of meeting with entrepreneurs, scientists, engineers and program managers from NASA and the DoD who were in awe of the breadth and depth of Mino's imagination and intelligence. He drew upon a wide range of concepts and technologies to invent new paradigms that challenged the status quo, sometimes shocking the existing order with the possibilities of his innovations.

Some skeptics and naysayers were quick to reject his ideas, and yet his underlying mathematics were tested and proven correct time and again. We travelled together across the USA, and found that our community is ready for these bold visions. I remember meeting with colleagues in the Pentagon who were fed up with the astronomical costs of conventional space systems and were delighted and excited to listen to his vision of applying nanotechnology to pressing current needs in space operations.

A. Weston
Director of Programs and Projects, NASA Ames Research Center, Moffett Field, CA, USA

F. Freund, S. Langhoff (eds.), *Universe of Scales: From Nanotechnology to Cosmology*,
Springer Proceedings in Physics 150, DOI 10.1007/978-3-319-02207-9_4,
© Springer International Publishing Switzerland 2014

It is a great tribute to Mino to see today how NASA and the national space community including the National Reconnaissance Office, DARPA, the Air Force and the Army have embraced his contributions and have begun to implement his vision.

Mino is certainly one of the key motivators and inventors of the ideas and technologies that have led to the explosion of cube-sats that have become a critical enabler to students around the world to gain experience and opportunity in space technology that has been denied them by the traditional approaches. When I look around Ames, NASA and the world today, there are thousands of researchers of all ages from around the world that have the opportunities to expand their horizons and make a real impact on space technology through small spacecraft that Mino pioneered. This lasting contribution is already resulting in a renaissance of enthusiasm and participation in space science and technology that had seemed increasingly out of reach in the last several decades.

Phone Call at 2 am

David Morse

Mino was a man of almost scary intelligence, boundless energy and amazing creativity and humor. He was brilliant and dedicated and hard working, but yet he also had a gift for having just the right touch to apply to everything he did. And although he would probably deny it, it wasn't truly a gift of course, it never is. It was something that Mino worked at diligently and with purpose, like every single thing he did in his short life.

At NASA I am what you call a "Global Approver," in our time card system. I check time cards all hours of the day and night, weekends, whenever. So one Saturday night—it was already early Sunday morning, I was up, unable to sleep (it happens as you get older), staring at my computer screen. "Let me check the time cards." Mino's wasn't done, no hours, nothing. I thought: "Well, I've got Mino's work phone number, I'll call his office and leave a message."

Mino answered the phone on the first ring—at about 2:15 am. When he answered, he said: "Good morning David, this is Mino." Well, it turned out that I'd called him once before, he'd bookmarked my number, and kept it. That was Mino, still up long past midnight, working. Of all the interactions I'd had with Mino, this one may be the silliest and the most trivial, and yet it's the one that persists in my memory.

D. Morse
NASA Ames Technology Partnerships Division, Moffett Field, CA, USA

F. Freund, S. Langhoff (eds.), *Universe of Scales: From Nanotechnology to Cosmology*,
Springer Proceedings in Physics 150, DOI 10.1007/978-3-319-02207-9_5,

Witness to a Formidable Dialog

Robert Dumais

I thought so highly of Mino. He epitomized the definition of a truly brilliant mind. One amazing attribute was his ability to communicate complex concepts in an effective and concise manner. He intrigued me with his depth of knowledge in all things Science, and I feel privileged to have had the opportunity to work with Mino.

I had set up a meeting at the Logyx corporate office on North Whisman Road in Mountain View, CA to discuss potential collaborations between NASA and the Department of Energy. The meeting included Richard Joseph, Ph.D., Mino Freund, Ph.D., and me.

Our discussions started as a business collaboration discussion and rapidly evolved into an in-depth scientific discussion between Mino and Dr. Joseph. I sat at the table in the Logyx conference room and almost immediately became a non-contributing spectator.

The communication evolved from a mere discussion into an impassioned scientific discourse. I was able to understand perhaps 25 % of the conversation as Dr. Joseph and Mino mentally sparred and intellectually fed off each other.

Much time has passed and I don't recall the details of the conversation, but I remember that the discussion was centered on new energy concepts.

Somewhere amidst the discussion, Dr. Joseph picked up the White Board marker and started putting a complex mathematical formula on the board. This formula was used as a mechanism to help further communicate the concept while the discourse continued. At one point the white board marker was set down as the mathematical formula took a secondary importance to the verbal communication.

As a result, the formula remained unfinished. Finally, after twenty or so minutes the two respected Scientists had satisfied themselves as their respective scientific points were made and the discussion slowly subsided. Without missing a beat, Mino in his understated way, walked over to the white board, picked up the marker, and completed the last line of the formula. Mino, with his easy manner, turned around

R. Dumais
President Logyx, LLC, Mountain View, CA, USA

F. Freund, S. Langhoff (eds.), *Universe of Scales: From Nanotechnology to Cosmology*,
Springer Proceedings in Physics 150, DOI 10.1007/978-3-319-02207-9_6,
© Springer International Publishing Switzerland 2014

with a smile of satisfaction. Both Mino and Dr. Joseph looked at the white board and nodded in approval with the finished formula.

I sat in deep appreciation of what I had just witnessed.

Mino erased the white board as if to indicate that the mental gymnastics were commonplace. We set up a follow-on meeting to continue the collaboration pursuit. Mino headed back to NASA for another meeting. Dr. Joseph readied himself to leave for the airport. Just before he left, he stated that Mino was truly a brilliant mind and one of the best and brightest rising stars at NASA. I agreed with Dr. Joseph and wished him well as he departed for the airport.

I remember sitting in the Logyx office after Mino and Dr. Joseph had left, thinking about how fortunate I was to witness such an exchange.

Mino was a brilliant mind and a brilliant light. The depth and breadth of his knowledge was impressive. He had an easy-going manner, coupled with a strong drive to contribute in a scientifically meaningful way to NASA and humankind.

I feel truly enriched for having known and worked with him.

Written with fond memories and the deepest respect and appreciation.

Stunning Intellect and Pure Mind

R. Norris Keeler

I first met Mino during one of the ERPS workshops at Scripps Institution of Oceanography in San Diego. Since Directed Technology, Inc. was the prime contractor on ERPS, and Friedemann (Mino's father) was involved, I suggested that Mino might visit DTI on his next travel to Washington, as he went there often on assignment for NASA Ames.

It was not much more than a month later that Mino showed up at DTI. He had the better part of the afternoon available, so we chatted for about an hour. Although I was not aware of it at that time, Mino had worked with a group that was interested in the Cosmic Microwave Background, and when I told him that the Cosmic Microwave Background spatial spectrum showed several decades of a $-11/3$ slope, he answered my question as to what that meant.

"Since you are mentioning this slope, that sounds like some of Kolmogorov's laws, and that means turbulence."

"So what does that mean from a cosmological standpoint?"

"Well," said Mino, "that means that about 300,000 years after the Big Bang, the universe was in turbulent motion."

"Do you know how stars are formed?"

Mino then described conventional views on this, that they are the result of matter pulled together by gravity, the attraction and impaction leading to thermonuclear initiation—a star is formed.

I asked Mino what role turbulence could play. He responded that "When the matter starts to come together, turbulence disperses it. This leaves baryonic mass there in space. These objects could become dark matter."

"So the observation of the turbulence spectrum gives us a clue to the formation of dark matter", I stated.

"Well", Mino said, "if neutrinos have a mass, however small, there are so many of them, they could also contribute to dark matter."

R. Norris Keeler
Director of Directed Technology, Inc., former Head of Physics at the Lawrence Livermore National Laboratory, McLean, VA, USA

F. Freund, S. Langhoff (eds.), *Universe of Scales: From Nanotechnology to Cosmology*,   15
Springer Proceedings in Physics 150, DOI 10.1007/978-3-319-02207-9_7,
© Springer International Publishing Switzerland 2014

"How could any of that influence dark energy?"

"Some of that baryonic mass could create an obscuration greater than expected", Mino replied, "but it is hard to see how neutrinos could have any effect on the ongoing observations indicating the presence of dark energy."

This discussion was on the cutting edge of current cosmology.

By that time, Ira Kuhn was available, so I told him that Mino was here. "He is my replacement", I said. Since I was getting on toward retirement, DTI was looking for a person to take over my role. Mino then joined us and discussed the programs he was putting together for Pete Worden at the NASA Ames Research Center.

It was amazing how much Ira and Mino were in synch. After Mino left, Ira commented: "It's obvious, this guy is a lot more than your replacement."

Ira offered Mino the office and clerical support of Directed Technologies, Inc., on his Washington visits, and in fact Mino returned several times during his Washington assignments, before his illness struck.

Mino and I were planning two scientific projects. The first was the determination of the metallization pressure of diamond. Mino could obtain made-to-order chemical vapor deposition blanks, and my Russian colleagues could shock them up to ~10 Megabars. The results could provide guidance to static worker, as to how far their diamond anvils could hold out, and provide upper limits on the diamond to metal phase transition.

Some time later, Mino pointed out to me that a new solid state gravimeter had been found that was several orders of magnitude more sensitive than current gravimeters, and was it possible it could be used to remotely detect the gravitational dipole of a submerged submarine? It was close to being possible, but before we could get started, Mino fell ill.

There was a certain purity about Mino's persona. Not only was he devoted to his parents, but he seemed to be totally devoid of all those vices, big and small, that seem to always be present in individuals that bright and accomplished. It would have been a delight working with him as a colleague—so I have lost not only a wonderful colleague, but also a lifelong friend. With his depth and breadth of intellect, he could appeal to every one of us, with his loss, we have all died a little inside.

Growing Up

Friedemann Freund

When Mino was barely three, I started to take him with me to my laboratory at the University of Göttingen in Germany. It was in the old Chemistry building, dating back to the late 1700s, next to the tree-filled park that encircles the old town. The ceilings were high, the wooden floor planks dark from thousand chemicals spilled by generations of students. I set up a low table for Mino in the corner of my second floor office with large windows and tree branches touching the glass panes. I gave Mino a prism the size of his little hand so that he could catch the specks of sunlight falling through the leaves outside the window. He saw how the sunlight spread into colorful bands. I showed him the spectra of an incandescent bulb, a fluorescent tube, a Bunsenburner flame, and a high voltage discharge through air—all distinctly different from the spectrum of sunlight.

Mino was fascinated by liquid nitrogen. He proudly donned oversize safety goggles and safety gloves. He learned how to handle the cold liquid nitrogen without hurting himself. He poured it onto the floor, watching with delight the wallowing white clouds of condensed moisture. He took flowers, which we had plucked in a meadow during our last stroll, dipped them into the dewar filled with liquid nitrogen, and shattered them like glass with a gentle hammer blow.

Years later, when Mino was working on his Ph.D. thesis in solid state physics at the ETH Zürich, he cooled samples with liquid helium, much colder than liquid nitrogen, only four degrees Kelvin above absolute zero, requiring much more elaborate cooling techniques.

One day in the mid-1980s, when I visited Zürich, Mino showed me a "toy", he had just finished building, a Scanning Tunneling Microscope. He told me about Heinrich Rohrer, a physicist at IBM Research Zürich, who was affiliated with the Physics Department at the ETH. Mino had worked with Heini Rohrer and Gerd Binning at the time of the development of the Scanning Tunneling Microscope. Mino decided to build one himself, and he did. He showed me the delicate mechanical

F. Freund
GeoCosmo Science Center, Los Altos, CA, USA

F. Freund, S. Langhoff (eds.), *Universe of Scales: From Nanotechnology to Cosmology*,
Springer Proceedings in Physics 150, DOI 10.1007/978-3-319-02207-9_8,
© Springer International Publishing Switzerland 2014

parts which must be stable enough to image individual atoms at a resolution of better than one Ångstrom, a ten millionth of a millimeter, and the electronic part, which uses intricate feedback loops to translate tiny electric currents into images.

Mino zeroed in the scanning microscope tip onto the test surface, a graphite single crystal, finely tuning a number of amplifier knobs in the electronics rack. A big smile rushed over his face when the hexagons of the graphite structure, made of six carbon atoms, appeared on the computer screen. I later heard that the Scanning Tunneling Microscope, which Mino had completed almost casually on the side and donated to his fellow ETH students in the Advanced Physics Lab, was the first ever built outside IBM, even before Binning and Rohrer published their design and were soon thereafter awarded the Nobel Prize in Physics.

Mino had golden hands. Whatever he touched, in the realm of science or the arts, came out perfect. While in high school he attended evening courses at the Music Conservatory studying piano and music theory. Then he became interested in ink drawing. Carrying his drawing pad wherever he was—Cologne, Zürich, Paris, New York, Florida, Japan—placing with a sure hand ink line after ink line on the paper, in a steady pace, never a wrong stroke, never the need for correction. For Mino it was a matter of fact that he would do things the best he could, and he cheerfully accepted that everything turned out astoundingly well.

During his Ph.D. thesis at the ETH in Zürich Mino worked with Alex K. Müller, who had just received the Nobel Prize for the discovery of the high Tc Cu-oxide based superconductors. This sparked Mino's longtime interest in electrical properties of oxides, which later turned out to be crucial for understanding many basic processes in the Earth and Planetary Sciences, in geophysics, astrophysics, and the evolution of the early Earth.

When Mino came to the University of California Berkeley as a Post-Doc in 1991, he was given the task to complete the construction of an ultralow temperature refrigerator to be used on a joint Japan-USA satellite mission, IRTS, Infrared Telescope in Space, designed to tally stars and galaxies and to peer back in time as far as possible.

Mino determined that the existing design, which had been put together by a previous post-doc, was flawed and would fail to reach sub-Kelvin temperatures in space. However, there were only eighteen months left until the launch of the IRTS satellite—considered too short to start all over with the design and construction of such a complex, centrally important device.

Mino's post-doc advisor, Andrew Lange, was the US PI for the IRTS mission. He was an ambitious assistant professor, brilliant in his field, expected by many to climb to the top of cosmology.

Andrew Lange panicked and put all blame for the failure on Mino.

Quietly, Mino set out to redesign the refrigerator. He built it new from scratch. He reached 260 milliKelvin—the design temperature needed to record the far infrared radiation coming from near the edge of the universe—light that has taken some ten billion years to reach our little corner in the Milky Way galaxy.

Mino tested—at NASA Ames—the survivability of the fridge and its electronics by putting them through grueling high vibrational loads mimicking those during launch, through extreme temperature cycles, high and ultrahigh vacuum exposures. The new fridge easily passed all tests.

In addition Mino obtained all technical approvals required by NASA and the Japanese Space Agency for every piece of hardware that goes into a satellite. He finished everything on time. IRTS launched on schedule.

IRTS was a highly successful mission, generating a wealth of data, which increased our understanding of the universe. For more than a decade Mino's refrigerator held the world record of the lowest temperatures achieved for the longest time in space, and IRTS became the only science satellite in the history of space that was ever recaptured by the Space Shuttle after completion of the mission and brought back to Earth. It is now on display in the Museum of Natural History in Ueno Park in Tokyo.

Andrew Lange never acknowledged Mino's contribution to the success of IRTS. True to his ambitious nature and hard-driving management style, he continued to make debasing remarks, taking all credit for himself.

When the news broke in January 2010 that Lange had committed suicide by hanging himself in a hotel room, Mino—already stricken by his deadly brain tumor—stared for a long time out the window. Then he said in a low voice: "Andrew was the closest to an emotionally empty man I have ever met. His cruelty was boundless, relentless, and—in the end—directed against himself."

After Japan and a full decade in cosmology Mino joined the NASA Goddard Space Flight Center and three years later the Air Force Research Laboratory in Dayton, OH. At NASA Goddard he worked on an early version of the exquisitely sensitive detector, HAWC, designed to go on SOFIA, NASA's Stratospheric Observatory for Infrared Astronomy, to image the sky in the far infrared region. According to Al Harper, the HAWC Principal Investigator, Mino's contributions came at a critical time during the early stages of the development. At AFRL Mino put together the nanotechnology portfolio for the entire Air Force.

When Mino came to the NASA Ames Research Center, he made nanotechnology the centerpiece of his activity and took this field further than anybody had ever thought of before. He saw enormous potential, not only in space applications and new materials but also in neuroscience.

Through 2008 and the first half of 2009 Mino had been busy setting up collaborations across the USA to bring to bear the power to nanotechnology in neuroscience, specifically for targeting brain tumors. For August 20, 2009, a Thursday, he had made arrangements to drive to Berkeley to meet a researcher at UCB and discuss a joint project on neuroscience nanotechnology.

But Mino did not feel well. He canceled the trip. He spent the next three days at home with a numbing headache. On Monday he drove to the Palo Alto Medical Foundation Emergency Room for an MRI. On the following day, Tuesday August 25, 2009, he was told that he had a brain tumor in his right parietal lobe, a GBM, the most aggressive form of a glioblastoma. It had already grown to the size of a lemon.

From that day onward nothing was as before.

Mino was fully aware that his chances for survival were minimal. With his characteristic determination he engaged in the battle and never lost hope.

Along the way he continued his work as much as he could. He finished his job as COTR and executive agent for the DARPA FAST program, which demonstrated an up to 60-fold improvement in the photovoltaic power generation, 130 W/kg as compared to the measly 3 W/kg on the ISS today. He co-wrote the final report by Boeing, co-authored the FAST Applications Study with Boeing, the Aerospace Corporation and AFRL.

In January 2011 NASA Ames Center announced the Center Innovation Fund opportunity. With only one day left to the deadline, already unable to use his left hand, Mino sat down at the computer and typed the proposal text in one sitting. It was ranked top among nearly one hundred submissions. This work produced fundamentally new insight into why rocks become softer when subjected to stress.

In the area of nano- and bio-nanotechnology Mino was expanding his activities. In May 2011, he participated in the Stanford Nano Probe Workshop, which resulted in several summer students coming to Ames. He organized and participated in an interim review of the DHS program at Ames on distributed trace gas sensors based on nanotechnology. He initiated contacts with the "Blue Brain" project at the Swiss Polytechnic Institute in Lausanne, Switzerland, which will operate within a few years at 3–5 Exa-flops, simulating brain functions down to the synapses level. He explored the possibility of a National Center for Bio-Nano-Electronics headquartered at Ames. He was on the Scientific Program Committee of the 2011 International Brain Mapping Conference in San Francisco and co-chaired its Space Medicine sessions. He received the "Beacon of Courage" for his perseverance in face of three near-death encounters. Mino wrote a White Paper to DARPA for renewed funding for nano-bio activities at Ames related to traumatic brain injuries.

In the area of earthquake early warning research, Mino had initiated—before he fell ill—a DARPA study to evaluate the infrared emission from the Earth surface prior to major seismic events. It involved NASA Ames and the Jet Propulsion Laboratory, JPL. Mino participated in the planning of a DHS Workshop "by invitation only" in Monterey on earthquake forecasting and response.

The last scientific meeting Mino was able to attend was a July 2011 workshop organized by Tom Jordan, Southern California Earthquake Center, on the state of the art of earthquake early warning and forecasting.

Mino had many plans for the future including his grandiose concept of a swarm of nanosatellites, all communicating with each other. They would, he predicted, enable new ways of worldwide communication. At NASA Ames he pushed for the development of nanosatellites at a time when almost everybody in aerospace thought that they were a futile dream. Some smiled and called them Mino-Sats. Today the Ames Research Center has become one of the hubs for this new technological breakthrough and dozens of groups around the world are working on Mino-Sats.

Part II
Science Chapters

Mino in Japan: The Infrared Telescope in Space

Michael D. Bicay

Abstract This Invited Lecture looks back twenty years and recalls the early professional career of Dr. Minoru (Mino) M. Freund, as seen through the eyes of the current Director for Science at NASA's Ames Research Center. The focus is placed on Dr. Freund's role on the *Infrared Telescope Space*, a joint Japan-NASA collaboration in the 1990s.

1 The "Decade of the Infrared" in Astronomy

I first became acquainted with Mino while serving as a Visiting Senior Scientist at NASA Headquarters (HQ) in the early 1990s. I had agreed to go to Washington, DC in late 1990 on a two-year appointment—and ended up staying for six years. I worked in the Infrared, Submillimeter and Radio Branch of the Astrophysics Division within the Office of Space Science and Applications.

Shortly after my arrival at NASA-HQ, the National Research Council's (NRC) Decade Survey for the 1990s, "*The Decade of Discovery in Astronomy and Astrophysics*," was published [1]. The top priority space-based mission was the *Space Infrared Telescope Facility* (*SIRTF*), the final element in NASA's Great Observatories program (which also included the *Hubble Space Telescope*). Another high priority in that guiding document was the *Stratospheric Observatory for Infrared Astronomy* (*SOFIA*), a next-generation airborne infrared telescope. The fact that these observatories rated high in the NRC Decade Survey led many to describe the 1990s as the "Decade of the Infrared" in astronomy and astrophysics.

With the wisdom afforded by hindsight, it is easy to understand why infrared astronomy was well positioned for expansion. Just as radio astronomy blossomed in the late 1940s and early 1950s, catalyzed by radar technologies developed and refined for military uses in World War II, infrared astronomy was enabled by the rapid advancement of detector technologies developed by the military for downward

M.D. Bicay (✉)
Director for Science, NASA Ames Research Center, Moffett Field, CA, USA
e-mail: Michael.D.Bicay@nasa.gov

F. Freund, S. Langhoff (eds.), *Universe of Scales: From Nanotechnology to Cosmology*,   23
Springer Proceedings in Physics 150, DOI 10.1007/978-3-319-02207-9_9,
© Springer International Publishing Switzerland 2014

looking applications. During my time at NASA-HQ, I funded a variety of university-based researchers to adapt sensors derived for military applications to use them for low-light astronomical applications. It would be these detectors that would form the basis of SIRTF, SOFIA and later infrared telescopes—on the ground, and in space.

2 Early Infrared Astronomy

Water vapor in Earth's atmosphere absorbs most of the infrared radiation we receive from the cosmos. To circumvent this fact of life, astronomers in the 1960s began attaching telescopes to huge balloons that would "launch" to altitudes of more than 30 km to gain a clearer view of the Universe. By the early 1970s, scientists attached small telescopes to poke out through ports in high-flying Lear jets and on sounding rockets to conduct measurements, thereby discovering thousands of new celestial sources. None of these telescopes could get completely above the atmosphere, however, and by the early 1970s astronomers began discussing the possibility of putting an infrared telescope in orbit. Most of the early plans involved carrying an IR telescope on repeated flights of NASA's Space Shuttle. I note that this idea was developed at a time when NASA thought that Shuttle would make routine flights nearly every week! More importantly, planning assumed that the contamination from the Shuttle (*e.g.*, vapors, small particulates, heat interference) would not be a significant problem.

In 1979, the idea for a *Shuttle Infrared Telescope Facility* (the "original" SIRTF) was highly recommended in a report published by the National Academy of Sciences. In 1983, NASA released a solicitation for science instruments for an infrared telescope that would remain attached to the Shuttle during its mission, and returned to the ground for refurbishment prior to its next flight. The first launch was anticipated around 1990.

As NASA made this announcement, the first infrared telescope was launched into space by a collaborative team from the United States, United Kingdom, and the Netherlands. The *Infrared Astronomical Satellite* (*IRAS*) was an Explorer-class satellite designed to conduct the first infrared survey of the celestial sky. [As an aside, the author used data from IRAS to research the infrared properties of galaxies in clusters, and to study the global correlation between thermal infrared emission and synchrotron radio emission from spiral galaxies.] In opening our eyes to the infrared universe, the 10-month IRAS mission was a resounding success [2, 3], and led to significant interest among astronomers for a follow-on mission.

The impressive scientific returns from the free-flying IRAS satellite led NASA to revise its announced plans for a Shuttle telescope "to provide flexibility for the possibility of a [free-flyer] SIRTF mission." In 1984, NASA selected a team of astronomers to build the instruments and define the science program for a free-flying, orbital version of SIRTF. Alas, it would be 2003 before SIRTF (later renamed the *Spitzer Space Telescope*) was launched into space—after two budget-driven redesigns and one Congressional cancellation!

3 Japan's Infrared Telescope in Space

Japan has enjoyed considerable success in launching astronomy telescopes aboard balloons, sounding rockets, and into space. In the 1990s, the Institute of Space and Astronautical Science (ISAS), a consortium of universities united to advance space research, administered these projects. [In 2003, ISAS would be merged with two other organizations to form the Japan Aerospace Exploration Agency, or JAXA.]

Japan's first space-borne infrared telescope was introduced in a paper presented at an international scientific conference in Dordrecht, the Netherlands in 1989 by Professor Haruyuki Okuda [4]. In this paper, Prof. Okuda described plans to attach a small, cryogenically cooled telescope to a deployable platform, and to conduct observations of the cosmic background radiation and Galactic infrared emission. At that time, launch was anticipated for 1994.

I was serving at NASA-HQ as the Agency's Program Scientist and Program Manager for what would become known as the *Infrared Telescope in Space* (*IRTS*). IRTS was comprised of a modest-sized 15 cm diameter telescope that was to be attached to the Space Flyer Unit (SFU). The SFU was a platform with seven science and technology experiments attached, and was the culmination of an inter-Agency effort involving ISAS, Japan's National Space Development Agency, and the Japanese Ministry of International Trade and Industry. The purpose of SFU was to provide researchers with the opportunity to fly experiments in space, using the US Space Shuttle to retrieve the SFU and return it to earth for data retrieval and analysis, and for possible refurbishment. I note that Mino was one of the 25 authors of the seminal pre-launch publication describing the IRTS mission [5]—and one of only five non-Japanese authors.

In the early 1990s, ISAS approached NASA seeking help to develop and operate IRTS. More specifically, NASA agreed to provide two of the four IRTS science instruments. The first was the Far-Infrared Photometer (FIRP), provided by the University of California at Berkeley. The FIRP [6] was designed to observe the Universe at four discrete wavelengths: 150, 250, 400 and 700 microns. At these far-infrared and sub-millimeter wavelengths, FIRP needed to be refrigerated to near absolute zero in order to reduce the ambient noise and allow IRTS to detect faint signals from interstellar dust and the very distant extragalactic background. FIRP had a spatial resolution of 0.5 degree, and a spectral resolution of $R = \lambda/\delta\lambda$ of 3 over a 30 arcmin square field of view (FOV).

The second NASA instrument was the Mid-Infrared Spectrometer (MIRS), designed and developed by Dr. Thomas (Tom) Roellig at NASA's Ames Research Center. The MIRS instrument [7] was one of three grating spectrometers on the IRTS payload, and covered the wavelength range of 4.5 to 11.7 microns using doped silicon detectors. MIRS provided a moderate spectral resolution of $R = 12$ to 51, and was designed to study the interstellar medium (ISM) and the infrared cirrus discovered by IRAS.

The other grating spectrometers aboard IRTS were the Near-Infrared Spectrometer (NIRS), built by Nagoya University, and the Far-Infrared Line Mapper (FILM), developed by ISAS and a consortium of Japanese universities. NIRS [8] used indium antimonide detectors to cover 1.4 to 4.0 microns, with $R = 11$–36, and was designed to study the zodiacal light from dust within our own Solar System, diffuse Galactic light from our Milky Way, and extragalactic sources. The FILM instrument [9] used stressed gallium-doped germanium detectors to provide coverage of the two primary cooling lines in the ISM: [O I] at 63 microns and [C II] at 158 microns. The FILM spectral resolution was $R = 130$ and 407. Each of these spectrometers featured a field of view of about 8 arcmin square.

While the spacecraft, SFU, IRTS telescope and science instruments were being developed, ISAS approached me at NASA Headquarters and inquired about the possibility of sending American scientists to Japan to help with operations and data analysis. And it is for this opportunity that I sought out the help of a young, talented and researcher who I quickly realized had enormous potential—and was fluent in Japanese—Dr. Mino Freund.

4 Mino's Role on IRTS

In addition to managing NASA's contributions to IRTS in the early 1990s, I also provided the same oversight for NASA involvement in a Japanese radio astronomy satellite later renamed the *Highly Advanced Laboratory for Communications and Astronomy*. I had come to understand that while the Japanese were technically very competent, they operated their space missions on a shoestring budget. Graduate students often provided much of the labor in conducting mission operations for Japanese astrophysics missions. Graduate students also performed most of the data processing and analysis for science instruments developed by university-based Principal Investigator teams. The latter was—and remains—common in the United States; the former is not.

At the time, Mino was a postdoctoral fellow at the University of California (UC) at Berkeley, under the direction of Professor Andrew E. Lange. I was providing technology and research funding to Prof. Lange and the leader of the long-wavelength instrument development group at Berkeley, Prof. Paul L. Richards. Both Richards and Lange were making significant progress in developing ultra-cold instruments to enable exploration of the far-infrared and sub-millimeter wavelengths regimes being opened by NASA's *Cosmic Background Explorer* (*COBE*) satellite. COBE was designed to measure the remnant radiation from the Big Bang and the far-infrared background light resulting from the superposition of known (and unknown) celestial sources. It was a spectacular success [10], symbolized by the fact that the 2006 Nobel Prize in Physics was awarded to two of the instrument Principal Investigators: John C. Mather and George F. Smoot. COBE provided conformation of standard cosmology and the primacy of the Big Bang model. While Mino had many scientific interests and published over a wide variety of disciplines through the course of his life, he always remained interested in cosmology.

Knowing that space-borne missions took more than a decade to come to fruition, Profs. Richards and Lange flew increasingly sensitive far-infrared instruments on sounding rockets and on balloons to follow-up the COBE results. Substantial progress by the Berkeley group (and others) led to large fractions of the celestial sky being mapped at sensitivities and spatial resolutions better than those achieved by COBE.

In the UC laboratories, Mino refined the instrument design and fabrication techniques he initially learned during his time as a physics doctoral student at the Swiss Federal Institute of Technology (or Eidgenossische Technische Hochschule, ETH) in Zurich in the mid-1980s. His training and expertise in low-temperature physics at ETH would be a huge asset and a saving grace during the development of the FIRP instrument for IRTS.

Before proceeding with my chronology, I must include an excerpt from a memorial narrative provided to me by Mino's father, Dr. Friedemann Freund, an accomplished Senior Research Scientist at the SETI Institute and NASA's Ames Research Center:

"[Already as a three-year-old in Germany] Mino was fascinated by liquid nitrogen. He proudly donned oversize safety goggles and safety gloves. He learned how to handle the very cold liquid nitrogen [77 Kelvin, or minus 196 degrees Celsius] without hurting himself. He poured it onto the floor, watching with delight the wallowing white clouds of condensed moisture. He took flowers, which we had plucked in a meadow during our last stroll, dipped them into the dewar filled with liquid nitrogen, and shattered them like glass with a gentle hammer blow."

During his time at ETH in Zurich, Mino worked routinely with liquid helium, a much more challenging cryogen capable of cooling devices to 4 Kelvin, or minus 269 degrees Celsius. Working with helium imposed far more challenges on instrument cooling techniques, and on safety.

While at Berkeley, Mino faced perhaps his biggest technical challenge to date. Because the FIRP instrument needed to operate at far-infrared wavelengths, it needed to operate at very cold temperatures to reduce the radiated heat originating from warm electronics and other instrument and telescope interfaces. To achieve its scientific objectives, FIRP needed to operate at temperatures of 300 milli-Kelvin. The challenge of building the required cryo-refrigerator is best summarized by an excerpt from reference [6]:

"The three main requirements for the FIRP Helium-3 refrigerator are: (i) the ability to condense Helium-3 and operate in zero-g[ravity] with minimal heating of the focal plane, (ii) a long holdtime and high duty cycle in orbit, and (iii) a self-contained and compact design that requires no mechanical penetration of the instrument cavity."

Within his first year at UC-Berkeley, Mino concluded that the original design for the FIRP refrigerator was flawed, and would likely not achieve the necessary sub-Kelvin operating temperatures on orbit. With only eighteen months to launch, Mino redesigned, fabricated and tested the entirely rebuilt Helium-3 refrigerator. Once flown aboard IRTS, the FIRP instrument achieved the first sub-Kelvin temperatures in orbit [11]. I do not feel that Mino received enough credit for his herculean work in salvaging the FIRP cooling system, which proved crucial to making IRTS work at the longest wavelengths.

The IRTS telescope was launched into low-Earth orbit from Tanegashima Space Center in Japan on 1995 March 18 aboard an H-2 rocket. Following an 11-day commissioning phase, the telescope began observing on March 29. For five weeks, IRTS surveyed about 2700 square degrees (nearly seven percent) of the celestial sky. Operations continued for about five weeks, until the Helium-3 cryogen was exhausted on 1995 April 24. The IRTS, still attached to the SFU platform, was retrieved by the Space Shuttle *Endeavour* in 1996 January, and returned to Earth. Today, IRTS is scattered among various sites in Japan. Some of the IRTS infrastructure can be seen at the national Science Museum in Tokyo's Ueno Park. The telescope mirror is housed at ISAS in Sagamihara, Japan. The FIRP refrigerator is at the California Institute of Technology in Pasadena, where Prof. Lange moved from Berkeley—before his tragic suicide in 2010.

5 Mino's Science Research with IRTS

By virtue of being a co-investigator on the FIRP instrument, Mino played a central role in analyzing and interpreting the far-IR and sub-millimeter data from IRTS. One of the first FIRP results published were observations of interstellar dust emission at Galactic latitudes $|b| < 3.5$ deg. Mino and his colleagues did what most astronomers do when handed data from multiple bands: they started plotting color-color diagrams to look for trends that could yield insight into physics [12]. They found that the 150/250 micron and the 250/400 micron brightness ratios of the interstellar dust emission showed a slight decrease with increasing latitude within the 3.5 degree band centered on the Galactic plane. This result was consistent with observations made earlier by COBE. However, the 250/700 micron ratio showed a much stronger dependence on Galactic latitude. This led the FIRP team to posit that there were two components of interstellar dust: warm ($\sim$40 K) and cold ($\sim$4–7 K). This added further confirmation of an emerging understanding that started with the interpretation of IRAS data. Moreover, the FIRP team concluded that the physical properties of the interstellar dust must vary significantly at low Galactic latitudes.

An interesting development in the post-operational phase of the IRTS mission is the extent to which Mino became a key NIRS science investigator even though he was not a team member prior to launch. Mino, working with Martin Cohen at UC-Berkeley, developed an important point-source extraction and calibration program for IRTS near-infrared sources [13]. Perhaps the most important result to come out of NIRS resulted from a search for the near-infrared extragalactic background light (EBL). The NIRS team successfully measured the spectra of the diffuse emission for a wide range of the sky. A preliminary version of their results was presented at a conference a year after the IRTS mission [14], but it would be nearly a decade before the definitive results were published in a peer-reviewed journal [15]. This is a testament to the difficulty of dis-entangling the faint EBL emission from the various foreground emissions: discrete galaxies, Galactic emission from the interstellar medium in our own Milky Way, and the zodiacal emission from dust within our Solar System.

The NIRS data had significantly smaller error bars than previous COBE measurements, and revealed a significant isotropic emission in the 1.4 to 4.0 micron regime. The spectrum was stellar-like, but showed a spectral jump from the optical observations of the extragalactic background light. The measured near-IR flux of $35 \ \mathrm{nWm^{-2} \, sr^{-1}}$ was too bright to be explained solely by the integrated light from faint galaxies. A two-point angular correlation analysis showed spatial fluctuations of a few degrees in the NIRS data. The NIRS team posited that the spectrum and brightness of the observed EBL emission could be explained by the redshifted ultraviolet emission from the first generation of Population III stars, which presumably caused the re-ionization of the Universe.

Mino also worked with other Japanese members of the IRTS team to use data from both the NIRS and MIRS spectrometers to study water vapor absorption in early M-type stars. M-type stars are cool and faint, but also the most common stars, comprising 76 % of the stellar population in the Solar neighborhood. Prior to the IRTS mission, astronomers had measured water absorption primarily in M6 or later type stars. But Mino and his co-authors discovered water vapor absorption in about ten percent of the early M-type stars. The observed spectral features could be fitted with model spectra with excitation temperatures of only 1000–1500 K, and water column densities of 5×10^{19} to $1 \times 10^{20} \ \mathrm{cm^{-2}}$. From these results, the team concluded that water molecules were present in a region of the stellar atmosphere above the photosphere. The observed correlation between the intensity of the H_2O absorption and the mid-IR excess implied that the extended atmosphere was physically connected to the mass loss from these early M stars.

By the time Mino had moved on to other intellectual pursuits, he had co-authored 26 IRTS-related papers from 1993 to 2005, including eleven in peer-reviewed journals.

6 Final Thoughts

Even before the final IRTS results were published, Mino had expanded his horizons and begun to explore new vistas. He would work as an astrophysicist at NASA's Goddard Space Flight Center from 1999–2002, developing new and sensitive detectors to image the celestial sky at far-infrared wavelengths. He spent 2002–2005 as a research physicist at the U.S. Air Force Research Laboratory in Dayton, Ohio, working to develop a nanotechnology research portfolio for the USAF. Finally, he would come home—to the Bay Area and to NASA's Ames Research Center—in 2006 to become Director of the Center for Nanotechnology and Advanced Aerospace Materials and Devices. Unfortunately, he was with us for too short a time following the discovery of a brain tumor on 2009 August 25.

With the strong and endearing support of his parents, Friedemann Freund and Hisako Matsubara, and of his colleagues and friends at Ames and around the world, Mino fought valiantly for more than two years, before succumbing on 2012 January 17, two weeks shy of his 50th birthday. During this difficult period, he maintained a vigorous fighting spirit, and uplifted many through his online blog "A Little Detour."

Let me close this remembrance by sharing with you the thoughts that came to me spontaneously at Mino's viewing after he had passed, inscribed in a book offered by Mino's loving parents, Friedemann and Hisako:

"To honor a true Renaissance man. Your enthusiasm, intellect and curiosity were admirable and infectious. You lived two lifetimes in your short time here. You'll be remembered by many for a long time."

It was not until Mino's passing that I fully appreciated his extraordinary breadth of talent across the sciences—and beyond. He was fascinated with science and technology, obviously—but also by history, literature, religion and civilization. He was proficient in English, German, French and Japanese. He learned to play piano at a young age, and was interested in the history of music and composition. He became fascinated, and more than proficient, in art. While growing up in Germany, he began to draw the cityscapes of Cologne and landscapes along the Rhine River. He ventured to Paris and translated the City of Light into a series of remarkable ink drawings. In honor of his special talents, I include two of his sketches to end this contribution: of the Notre Dame Cathedral in Paris, and the Kenkun Shrine in Kyoto, Japan.

References

1. National Academies Press, *The Decade of Discovery in Astronomy and Astrophysics* (1991). Available at http://www.nap.edu/catalog.php?record_id=1634
2. G. Neugebauer et al., The Infrared Astronomical Satellite (IRAS) mission. Astrophys. J. **278**, L1 (1984)
3. G. Neugebauer et al., Early results from the infrared astronomical satellite. Science **224**, 14 (1984)
4. H. Okuda, Observations of diffuse infrared radiation by a small cryogenically cooled telescope, IRTS, in *Proceedings of a Conference held in Dordrecht, Netherlands* (Kluwer Academic, Dordrecht, 1990), p. 435
5. H. Murakami et al., The Infrared Telescope in Space (IRTS). Astrophys. J. **428**, 354 (1994)
6. A.E. Lange et al., The far-infrared photometer on the infrared telescope in space. Astrophys. J. **428**, 384 (1994)
7. T.L. Roellig et al., The mid-infrared spectrometer on the Infrared Telescope in Space (IRTS) mission. Astrophys. J. **428**, 370 (1994)
8. M. Noda et al., Near-infrared spectrometer on the infrared telescope in space. Astrophys. J. **428**, 363 (1994)
9. H. Shibai et al., Far-Infrared Line Mapper (FILM) on the infrared telescope in space. Astrophys. J. **428**, 377 (1994)

10. J.C. Mather et al., A preliminary measurement of the cosmic microwave background spectrum by the cosmic background explorer (COBE) satellite. Astrophys. J. **354**, L37 (1990)
11. M.M. Freund et al., Design and flight performance of a space-borne helium-3 refrigerator for the infrared telescope in space. Cryogenics **38**, 435 (1998)
12. T. Hirao et al., Submillimeter observations of the galactic plane by IRTS. Publ. Astron. Soc. Jpn. **48**, L77 (1996)
13. M.M. Freund et al., The NIRS point source extraction program and stellar catalogues. ASP Conf. Ser. **124**, 114 (1997)
14. T. Matsumoto et al., IRTS observations of the near-infrared diffuse background. ASP Conf. Ser. **124**, 334 (1997)
15. T. Matsumoto et al., Infrared telescope in space observations of the near-infrared extragalactic background light. Astrophys. J. **626**, 31 (2005)

Mino's Sense of Stardust

Yvonne Pendleton

Brilliance, kindness, and intellectual honesty—these are qualities that best define Mino Freund. They made him a wonderful person to know, a friend, and propelled him, on several occasions, to greatness. Those of us fortunate enough to have met him, will long miss him. My own research, on the origin and evolution of organic material in the interstellar medium, was one of the many fields that fascinated Mino. In 2006, he and his father, Friedemann Freund, co-authored a paper that contributed unique ideas and a fresh look at some aspects of the field. I remember the conversations we had that started his probing mind down the stardust path.

Interstellar dust is a ubiquitous component of our Galaxy, and Mino was intrigued. He wanted to understand the evolution of components found in that dust, knowing that its organic components may have survived the accretion of the planetary system and may have played a role in starting life in our quiet backwaters of the interstellar medium.

Infrared (IR) spectroscopy is a powerful tool with which we can study the interstellar dust. Systematic IR observations of different interstellar regions have revealed a rich and varied composition of dust grains [12]. There is a clear dichotomy of the IR spectroscopy of dust in diffuse and in dense regions of the interstellar medium, and yet we find remnants in our Solar System today that were formed in regions beyond dense cloud environments. So it seems the two are somehow linked through the dust evolution.

The two environments differ in a number of important ways such as gas density, gas temperature and composition, and intensity of the UV radiation field. Dense molecular clouds allow the formation of ice mantles, primarily composed of water, on dust grains. Surrounding the molecular clouds is the diffuse interstellar medium (DISM). Two prominent IR features are associated with mineral grains in the DISM: a 9.7 µm [1030 cm^{-1}] mineral, i.e. silicate signature and an "organic" signature [5] in the C–H stretch region around 3.4 µm [2920 cm^{-1}], with sub-features at 3.38,

Y. Pendleton (✉)
Director, NASA Lunar Science Institute, NASA Ames Research Center, Moffett Field, CA, USA
e-mail: Yvonne.Pendleton@nasa.gov

F. Freund, S. Langhoff (eds.), *Universe of Scales: From Nanotechnology to Cosmology*,
Springer Proceedings in Physics 150, DOI 10.1007/978-3-319-02207-9_10,
© Springer International Publishing Switzerland 2014

3.41, and 3.50 μm [2858, 2932, and 2857 cm^{-1}] [8]. The near-IR absorption bands are due to, respectively, the stretching and bending modes in the methyl (CH_3) and methylene (CH_2) aliphatic groups (e.g., [9, 10]).

While the aliphatic component is ubiquitous in the diffuse ISM, it has never been observed in dense molecular cloud materials. In the case of dense clouds, the C–H stretching region is generally characterized by absorption bands at 3.25, 3.47, and 3.54 μm [3077, 2882, and 2825 cm^{-1}]. A 3.47 μm band is a common feature of young stellar objects embedded in molecular clouds in quiescent regions [1, 3, 6, 11], and has been attributed to simple hydrocarbons. However, in dense clouds, the feature is significantly different in band peak position and band shape from the DISM hydrocarbon bands. These spectroscopic differences between dense and diffuse media provide a basic constraint on modeling the formation and evolution of the interstellar aliphatic dust grain component [2, 7].

Mino wondered, along with many of us, what paths the mineral and organic components followed after their creation in the interstellar medium (ISM). He wondered what could account for the observed differences between the dust in diffuse and dense clouds. There is ample evidence that interstellar grains are condensing in the outflow of post-main-sequence stars with photospheric temperatures between 2500 and 3500 K. This includes red giants, red supergiant and asymptotic giant branch stars. The dust created around these stars is transported into the low-density environment of the ISM via stellar mass loss, novae and supernova explosions. Low-density environments of the ISM are permeated by dissociative ultraviolet radiation and traversed by shocks, both capable of destroying dust grains.

Interstellar grains migrate from the diffuse medium into the dense medium of the ISM (dark globules and dense molecular clouds) where they encounter conditions that favor grain aggregation. In these dense clouds, dust grains are shielded from dissociative ultraviolet and suffer less shocks. Under these conditions rapid grain growth can occur due to aggregation and coagulation and volatiles can condense on the grains forming icy mantles. The grain mantles become a repository for atoms and molecules out of the gas phase, leading the way to molecular species such as H_2O, CH_3OH, CO and CO_2, all abundant constituents in the icy mantles, formed through surface chemical reactions. Over time the exchange of matter between gas and dust regulates the chemical evolution of the cloud as the grain mantles provide both sinks and sources for gaseous molecules.

By the time star formation is initiated and progresses, the dust grains and their mantles are further processed by stellar radiation and by winds from the newly born stars leading to chemical reactions, which drive the composition of the organics towards greater molecular complexity. Complex organic molecules such as dimethyl ether and methyl formate are abundant in the warm dense gas regions associated with newly formed stars. They are generally thought to form from processed and sublimated interstellar ices [4]. As protostars evolve, strong stellar winds sweep up and 'clear' their surroundings, culminating in the formation of cavities whose inner walls become prominent IR reflection nebulae. As a result, dust grains are deeply involved in the origin and evolution of molecular complexity of the organics in space, particularly in regions of star and planet formation. Various processes govern the interaction between gas and dust grains. They ultimately control the composition and molecular complexity of the matter from which planets are forming around newly born stars. Understanding these processes is crucial to the search for our origins.

The paper "Solid solution model for interstellar dust grains and their organics" by Mino Freund and his father, Friedemann Freund, published in March 2006 in the Astrophysical Journal [5], shines a very different light on issues related to the composition, structure, and evolution of interstellar dust grains than had been previously discussed in the astronomical literature.

The Freund & Freund ApJ paper [5] points to the formation of solid solutions between the refractory mineral phases and the gaseous compounds such as H_2O, CO and CO_2—an aspect largely overlooked by the astronomy and astrochemistry communities. In the process of forming solid solutions H_2O molecules are split into molecular H_2 inside the dust grains and molecules such as CO and CO_2 are converted into chemically reduced, i.e. "organic" carbon. Through solid-state processes well-known in the technical world such as supersaturation, segregation and exsolution, the reduced carbon and the H_2 molecules, both encapsulated inside the solid matrix of the dust grains, are made to react with each other, forming aliphatic hydrocarbons, which are protected from the harsh DISM environment by the mineral phase in which they reside. This fundamental process is capable of driving a rich organic chemistry.

The Freund & Freund ApJ paper [5] provides a new interpretation of key issues related to interstellar dust. It points to the possibility that mineral grains in the interstellar medium are deeply intertwined with their organics and are indistinguishable from them. It treats the nanometer-sized grains, which condense in the gas-rich outflow of late-stage stars or expanding gas shells of supernova explosions, not as mineral cores made of refractory, high melting point oxides or silicates but from the outset as solid solutions between the mineral phases and the gas-phase components H_2O, CO, and CO_2 that are omnipresent in environments where grains condense. Through a series of thermodynamically well-understood solid-state processes, these solid solutions may become "parents" of organic matter that segregate and exsolve inside the dust grains. Thus, the mineral dust grains and their organics become part of the same thermodynamically defined solid phase and, hence, physically inseparable.

Based on Mino's and Friedemann's deep understanding of solid state chemistry, "old" problems within studies of interstellar dust are re-examined and a fresh explanation is offered. The part that is left unspoken in their paper is also quite interesting: solid-state chemistry can facilitate complex reaction routes and a variety of organic substances may be synthesized along the way. It would be interesting to further develop this idea, identify the chemical routes involved, the molecules made, quantify the efficiencies of these routes and address their prebiotic relevance for the origin of life in newly forming planetary systems including the solar nebula and the Earth. Indeed, it has always been difficult to understand the origin and inventory of organic material for the rocky planets in the habitable zone of the Solar System, because the main reservoirs of carbon, the highly volatile species such as CO and CH_4, are normally thought to be excluded during formation. The solid solution chemistry "unearthed" by Freund and Freund provides a hitherto unexplored route to assess and study this reservoir of organics [5].

As with so many papers Mino Freund wrote and research areas he explored, this paper proposes novel ideas that would have benefitted from the many years he should have had ahead of him. One can only hope that other bright minds will pick up where he left off, and we can all remember that it was his light that first shone upon this path.

References

1. L.J. Allamandola, S.A. Sandford, A.G.G.M. Tielens, T.M. Herbst, Astrophys. J. **399**, 134 (1992)
2. L.J. Allamandola, S.A. Sandford, A.G.G.M. Tielens, T.M. Herbst, Science **260**, 64 (1993)
3. T.Y. Brooke, K. Sellgren, R.G. Smith, Astrophys. J. **459**, 209 (1996)
4. P. Ehrenfreund, S. Charnley, Annu. Rev. Astron. Astrophys. **38**, 427 (2000)
5. M.M. Freund, F.T. Freund, Astrophys. J. **639**, 210 (2006)
6. J.E. Chiar, A.J. Adamson, D.C.B. Whittet, Astrophys. J. **472**, 665 (1996)
7. V. Mennella, G.M. MuñozCaro, R. Ruiterkamp, W.A. Schutte, J.M. Greenberg, J.R. Brucato, L. Colangeli, Astron. Astrophys. **367**, 355 (2001)
8. Y.J. Pendleton, L.J. Allamandola, Astrophys. J. **138**, 75 (2002)
9. Y.J. Pendleton, S.A. Sandford, L.J. Allamandola, A.G.G.M. Tielens, K. Sellgren, Astrophys. J. **437**, 683 (1994)
10. S.A. Sandford, L.J. Allamandola, A.G.G.M. Tielens, K. Sellgren, M. Tapia, Y. Pendleton, Astrophys. J. **371**, 607 (1991)
11. K. Sellgren, R.G. Smith, T.Y. Brooke, Astrophys. J. **433**, 179 (1994)
12. A.G.G.M. Tielens, *The Physics and Chemistry of the Interstellar Medium* (Cambridge Univ. Press, Cambridge, 2005)

How Brains Make Decisions

V.I. Yukalov and D. Sornette

Abstract This chapter, dedicated to the memory of Mino Freund, summarizes the Quantum Decision Theory (QDT) that we have developed in a series of publications since 2008. We formulate a general mathematical scheme of how decisions are taken, using the point of view of psychological and cognitive sciences, without touching physiological aspects. The basic principles of how intelligence acts are discussed. The human brain processes involved in decisions are argued to be principally different from straightforward computer operations. The difference lies in the conscious-subconscious duality of the decision making process and the role of emotions that compete with utility optimization. The most general approach for characterizing the process of decision making, taking into account the conscious-subconscious duality, uses the framework of functional analysis in Hilbert spaces, similarly to that used in the quantum theory of measurements. This does not imply that the brain is a quantum system, but just allows for the simplest and most general extension of classical decision theory. The resulting theory of quantum decision making, based on the rules of quantum measurements, solves all paradoxes of classical decision making, allowing for quantitative predictions that are in excellent

---

I met Mino in October 2011 in Zurich, at the end of his last trip seeking medical treatment in Europe. What was supposed to be a relaxed family lunch turned into a passionate 3 hour long in-depth exchange spanning a wide range of deep scientific interests... and we talked about the topic presented in this Chapter. Mino's quick and sharp mind immediately grasped the essence of applying quantum concepts to processes taking place in the brain. He saw this as a pivotal step towards understanding how the brain works and how thoughts are formed. Mino and I were resonating with the excitement of powerful new ideas, with the same enthusiasm for science and for the power of transdisciplinary thinking. I had found a remarkable soulmate.—Didier Sornette

---

V.I. Yukalov (✉) · D. Sornette
Department of Management, Technology and Economics, ETH Zürich, Swiss Federal Institute of Technology, Zürich 8092, Switzerland

V.I. Yukalov
Bogolubov Laboratory of Theoretical Physics, Joint Institute for Nuclear Research, Dubna 141980, Russia

D. Sornette
Swiss Finance Institute, c/o University of Geneva, 40 blvd. Du Pont d'Arve, 1211 Geneva 4, Switzerland

F. Freund, S. Langhoff (eds.), *Universe of Scales: From Nanotechnology to Cosmology*, Springer Proceedings in Physics 150, DOI 10.1007/978-3-319-02207-9_11,
© Springer International Publishing Switzerland 2014

agreement with experiments. Finally, we provide a novel application by comparing the predictions of QDT with experiments on the prisoner dilemma game. The developed theory can serve as a guide for creating artificial intelligence acting by quantum rules.

1 What Is Brain Intelligence

The brain is the center of the nervous system in all vertebrates and most invertebrates. Only a few invertebrates, such as sponges, jellyfish, sea squirts, and starfish do not have one, though they have diffuse neural tissue. The brain of a vertebrate is the most complex organ of its body. In a typical human, the cerebral cortex is estimated to contain 86 ± 8 billion neurons [1], each connected by synapses to several thousand other neurons.

The functioning of the brain can be considered from two different perspectives, physiological and psychological. We do not touch here the physiological side of the problem that is studied in neurobiology, medicine, and is also modeled by neuron networks [2–4]. Our aim is to model the functioning of the psychological brain, which is studied in cognitive sciences.

The ability of the brain to take decisions is termed intelligence. There exist numerous and rather lengthy discussions attempting to describe what intelligence is [5–12]. Summarizing these discussions, the basic feature of intelligence, which can be accepted as its brief definition, is *the ability of adaptation to the environment by the process of making optimal decisions*. This implies that the notion of intelligence is foremost the ability of making decisions. It is generally accepted that humans possess the highest level of intelligence in the animal kingdom. But animals also are able to take decisions, to adapt to their environment and to solve problems [13]. Thence, animals also possess intelligence. This concerns all animals, such as birds, fish, reptiles, amphibians, and insects. Moreover, other living beings, say plants, in some sense, do adapt to surrounding by making decisions [14]. Therefore, we need to accept that, to some degree, all alive beings have a kind of intelligence, since all of them adjust to their environment by reacting to external signals. Thus, one can talk of the intelligence of plants, fungi, bacteria, protista, amoebae, algae, and so on. In that sense, any entity that is able to take decisions, adapting to surrounding signals, can be assumed to have something like intelligence. If such an entity that is able to take optimal decisions is created by humans, it is called artificial intelligence [15].

In the following, we shall be mostly concerned with the functioning of the human brain, though many parts of our considerations could be applied to the functioning of the brains and nervous systems of other alive beings. The human brain, being the most developed and complex, exhibits in the most explicit way the features that could be met in the behavior of other animals. The aim of this paper is to demonstrate that the human brain makes decisions in a rather intricate way that cannot be described by the classical utility or prospect theory used in economics. We argue that decisions made by brains are not the same as straightforward computer-like calculations. Human decisions are based on the functioning of and interplay between conscious as well as subconscious processes of the brain. This complex behavior can be represented by the techniques of quantum theory, which seems to be the most general and simplest framework for realistically characterizing the decision making process of human brains.

The plan of the paper is as follows. In Sect. 2, we recall how decisions are supposed to be made by fully rational decision makers who evaluate the utility of prospects and choose the one with the largest utility. Such a strictly deterministic behavior is a strong simplification of the reality. Empirical observations show that there always exists a distribution of choices made by different subjects, rather a single optimal behavior. Even the same subject, under varying conditions or time, can make different choices when confronted with the same set of competing prospects.

This implies that the first step towards a realistic representation of decision making is the reformulation of classical utility theory within a probabilistic framework, which is accomplished in Sect. 3. Analyzing the signals, the subject formulates a set of possible actions, π_j, termed prospects that are weighted with probabilities $p(\pi_j)$. Taking a decision means the selection of an optimal prospect π_* characterized by the largest probability, though other prospects can also be chosen, with lower probabilities, that is, with lower frequency. The possible actions are always weighted with a probability distribution. This describes the probability weighted diversity of choices among a population of similar decision makers. There always exists a probability that some of the members choose one prospect, while others choose another prospect, although the majority prefers the optimal prospect. This is the essence of the probability weight that is associated with the frequentist interpretation, which defines the fraction of those who choose the related prospect.

Although the probabilistic utility theory that we introduce in Sect. 3 generalizes the standard deterministic utility theory, it does not take into account that real decision makers are not fully rational. Moreover, they experience a variety of emotions and behavioral biases. As a result, decisions are taken not by a simple evaluation of utilities but are essentially influenced by these biases and emotions. In taking decisions, two brain processes are involved, conscious and subconscious. This dual functioning of the brain makes its principally distinct from the straightforward calculations by a computer, as is discussed in Sect. 4.

To take into account this complex dual behavior, Sect. 5 presents a generalization of decision theory, which invokes the techniques of the quantum theory of measurements. The developed Quantum Decision Theory (QDT) contains none of the paradoxes that are so numerous in classical decision making. Importantly, we show that classical decision theory constitutes a particular case of QDT. The latter reduces to the former under a process that can be called "decoherence", which describes how the addition of reliable information decreases the emotional component of a decision, thus making it more and more controlled by the rational utility component.

To illustrate how QDT describes how decisions are made, avoiding the paradoxes of classical decision making and providing quantitative predictions, we treat in Sect. 6 the prisoner dilemma game.

Section 7 summarizes the results, stressing that the developed QDT is, to our knowledge, the sole decision theory that not merely removes classical paradoxes, but provides *quantitative* predictions, with *no adjustable parameters*, which are in good agreement with empirical observations.

Concluding this introduction, our main hypothesis is that the brain makes decisions through a procedure that is similar to quantum measurements. This does not require the brain to be a quantum object, but merely takes into account the dual nature of the decision process, involving both conscious logical evaluations as well as subconscious intuitive feelings. This chapter summarizes the Quantum Decision Theory (QDT) that we have developed in a series of publications [16–21]. We also provide a novel application on the prisoner dilemma game, comparing the predictions of QDT with experiments.

2 Choosing a Prospect on Fully Rational Grounds

Assuming that the subject is fully rational and possesses the whole necessary information for making decisions, it is reasonable to suppose that such decisions are based on the evaluation of the utility of the results following the corresponding action. This is the central assumption of expected utility theory, which prescribes a normative framework on how decisions are made. The basic mathematical rules of expected utility theory have been compiled by von Neumann and Morgenstern [22] and Savage [23]. Below, we give a brief sketch of the main features of utility theory in order to introduce the terminology to be used in the following sections, where the generalizations of this theory will be considered.

The outcomes of actions, that is, the consequences of events, are measured by payoffs composing a set

$$\mathbb{X} \equiv \{x_n \in \mathbb{R} : n = 1, 2, \ldots, N_{out}\}. \tag{1}$$

The number of outcomes N_{out} can be as small as two or asymptotically large. Positive outcomes correspond to gains, while negative ones to losses. Payoffs x_n can come with different probabilities $p_j(x_n)$, being labeled by an index $j = 1, 2, \ldots, L$, and satisfying the normalization condition

$$\sum_{n=1}^{N_{out}} p_j(x_n) = 1, \quad 0 \le p_j(x_n) \le 1. \tag{2}$$

The ensemble of payoffs and their probabilities is called a lottery, or a prospect

$$\pi_j = \{x_n, p_j(x_n) : n = 1, 2, \ldots, N_{out}\}. \tag{3}$$

One also uses the notion of compound lotteries that are the linear combinations of a given set of lotteries, with the same payoffs and with the linear combinations of the related weights.

There can exist several prospects forming a family

$$\mathcal{L} = \{\pi_j : j = 1, 2, \ldots, L\}. \tag{4}$$

The task of decision making is to decide between the prospects π_j, choosing one out of the given family.

The choice involves the classification of outcomes according to their utility for the decision maker. One defines a utility function $u(x) : \mathbb{X} \to \mathbb{R}$ that can also be called pleasure function, satisfaction function, or profit function. By definition, the utility function is nondecreasing (more is always preferred), so that $u(x_1) \geq u(x_2)$ for $x_1 \geq x_2$ and concave (diminishing marginal utility), such that $u(\alpha_1 x_1 + \alpha_2 x_2) \geq \alpha_1 u(x_1) + \alpha_2 u(x_2)$ for non-negative α_i's normalized to one. The first derivative $u'(x) \equiv du(x)/dx$ is termed the marginal utility that is non-negative for a non-decreasing function. The second derivative $u''(x) \equiv d^2 u(x)/dx^2$ is non-positive for a concave function. Hence, the marginal utility $u'(x)$ does not increase. This implies that, with increasing payoff x, the utility function decelerates. Such a function is termed risk averse [24, 25], since a sure payoff is always preferred to different random payoffs with the same mean value. The risk aversion can be captured by the so-called degree of risk aversion $r(x) \equiv -u''(x)/u'(x)$, which is non-negative. Examples of utility functions are linear, power-law, logarithmic or exponential functions. Usually, the utility of nothing is set to zero, $u(0) = 0$, but the absolute utility level is inconsequential.

Generally, a payoff x_n can be either positive, representing a gain, or negative corresponding to a loss. Strictly speaking, it is impossible to lose something, while having nothing. Even the poorest person can lose a gamble and go in debt, having an instantaneous negative net worth. However, taking into account the value of future incomes gives in general a positive net value and the debt then constitutes a loss of a part of future incomes. There can be however situations where debt reaches levels beyond the most optimistic expectations of future incomes, so that one has lost what one did not own now or will ever have in the future. In its encyclopedic review of the history of debt in human societies, Graeber documents that such situations were quite common [26]. They were usually followed by slavery (and are still in various explicit or disguised forms followed by some kind of slavery), where the person in debt sells his children, wife or himself. A loss is then backed up by the ultimate reservoir of wealth, being stored in the value of oneself [26]. Formally, this implies that, before losing x_n, one has an initial given amount $x_0 \geq x_n$. Then, shifting all payoffs by x_0, one can redefine the lottery so that all its payoffs be non-negative.

Each prospect is characterized by the expected utility

$$U(\pi_j) = \sum_{n=1}^{N_{out}} u(x_n) p_j(x_n). \tag{5}$$

This notion was introduced by Bernoulli [27] and an axiomatic theory was developed by von Neumann and Morgenstern [22], where the payoff weights were treated as objective. Savage [23] extended the notion to subjective probabilities evaluated by decision makers.

Expected utility can be interpreted either as a cardinal or ordinal quantity. Cardinal utility is assumed to be precisely measured and the magnitude of the measurement is meaningful. It can be measured in some chosen units, similarly to how distance is measured in meters, or time in hours, or weight in kilograms. However, such a definition in precise units is often impossible and, actually, not necessary. It is sufficient to interpret the expected utility as ordinal utility, for which its precise magnitude is not important, but the magnitude of the ratios between different utilities carries sufficient meaning.

The prospect uncertainty is described by the prospect variance

$$\text{var}(\pi_j) \equiv \frac{1}{N_{out}} \sum_{n=1}^{N_{out}} \left\{ x_n^2 p_j(x_n) - \left[\overline{x}(\pi_j) \right]^2 \right\}, \tag{6}$$

whose square root can also be called the prospect volatility or spread. We have used the following notation for the prospect mean

$$\overline{x}(\pi_j) \equiv \frac{1}{N_{out}} \sum_{n=1}^{N_{out}} x_n p_j(x_n) . \tag{7}$$

The ordering of the prospects is prescribed by the relations between their expected utilities. One says that a prospect is preferable to another one if its utility is larger than that of the latter. Two prospects are termed indifferent when their utilities coincide. The properties of the utility function $u(x)$ prescribe the properties of the expected utility.

(i) *Completeness*: For any two prospects π_1 and π_2, one of the following relations necessarily holds, either $\pi_1 = \pi_2$, or $\pi_1 < \pi_2$, or $\pi_1 > \pi_2$, or $\pi_1 \le \pi_2$, or $\pi_1 \ge \pi_2$, understood as the corresponding relations between their expected utilities.

(ii) *Transitivity*: For any three prospects, such that $\pi_1 \le \pi_2$ and $\pi_2 \le \pi_3$, it follows that $\pi_1 \le \pi_3$.

(iii) *Continuity*: For any three prospects ordered so that $\pi_1 \le \pi_2 \le \pi_3$, there exists $\alpha \in [0, 1]$ such that $\pi_2 = \alpha \pi_1 + (1 - \alpha)\pi_3$.

(iv) *Independence*: For any $\pi_1 \le \pi_2$ and an arbitrary π_3, there exists $\alpha \in [0, 1]$ such that $\alpha \pi_1 + (1 - \alpha)\pi_3 \le \alpha \pi_2 + (1 - \alpha)\pi_3$.

The central aim of expected utility theory is to calculate the expected utilities for all prospects from the given family $\mathcal{L}$ and, comparing their values, to find the prospect possessing the largest utility. Then the decision is taken by selecting this prospect corresponding to the largest utility, which is called the *most useful prospect*. The decision making scheme based on expected utility theory is given in Fig. 1.

3 Probabilistic Approach to Expected Utility Theory

According to the expected utility theory delineated above, the choice of a prospect is with certainty prescribed by the utility of the prospects. This theory is deterministic, since the choice, with probability one, is required to correspond to the prospect with the largest expected utility. Such a completely deterministic formulation contradicts the known empirical facts demonstrating that, under the same conditions,

Fig. 1 Scheme of the deterministic decision making process based on the choice of the most useful prospect with the largest expected utility

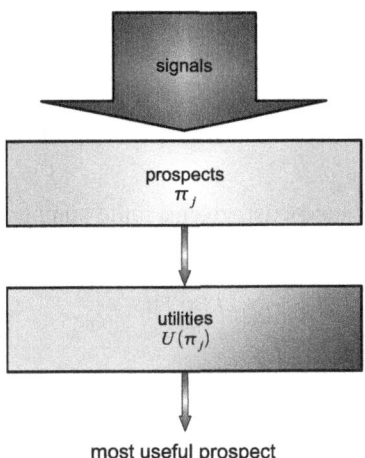

different persons often choose different prospects. Of course, one could salvage the deterministic theory by introducing heterogenous utility functions that describe the variety of tastes of different people [28]. While this captures the evident observation that tastes exhibit some heterogeneity, extending utility theory to heterogeneous or random utility theory comes at the cost of a proliferation of parameters, making the approach descriptive at best, while being non-parsimonious and non predictive. An even more convincing attack to the deterministic approach comes from the observation that the same person, under the same conditions, may choose different prospects at different times. This "intra-observer variation" has been largely documented in the medical literature [29, 30]. This suggests to view the brain of a decision maker as deliberating on the set of admissible prospects and evaluating them by involving some probabilistic weighting. This is the motivation to reformulate utility theory by generalizing it to a probabilistic approach.

The probabilistic weighting of prospects can be formalized by invoking the principle of minimal information that allows one to find a probability distribution under the minimal given information. The idea of this principle goes back to Gibbs [31–33], who formulated it as a conditional maximization of entropy under the given set of constraints. This principle is widely used in information science [34] and in physics [35, 36]. A general convenient form of an information functional is given by the Kullback-Leibler relative information [37, 38].

In order to weight the prospects according to their utility, let us consider a family of prospects $\mathcal{L}$. Assume that they can be weighted by means of a distribution defined by *utility factors* $f(\pi_j)$ that are normalized,

$$\sum_{j=1}^{L} f(\pi_j) = 1, \quad 0 \leq f(\pi_j) \leq 1. \tag{8}$$

By definition, the utility factor of zero utility is to be zero,

$$f(\pi_j) = 0, \qquad U(\pi_j) = 0. \tag{9}$$

Since the utility factors $f(\pi_j)$ weight the finite utilities $U(\pi_j)$, the total finite expected utility defined by

$$U = \sum_{j=1}^{L} U(\pi_j) f(\pi_j) \tag{10}$$

should exist, given a finite number L of prospects.

Under these conditions, we can define the Kullback-Leibler information as

$$I[f] = \sum_j f(\pi_j) \ln \frac{f(\pi_j)}{f_0(\pi_j)} + \lambda \left[\sum_j f(\pi_j) - 1 \right] - \beta \left[\sum_j U(\pi_j) f(\pi_j) - U \right], \tag{11}$$

with a trial distribution $f_0(\pi_j)$ proportional to the expected utility $U(\pi_j)$ in order to take into account condition (9). The parameters λ and β are the Lagrange multipliers guaranteeing the validity of the imposed constraints (normalization (8) and existence of a well-defined finite expected utility (10)).

Minimizing the information functional (11) yields the utility factor

$$f(\pi_j) = \frac{U(\pi_j)}{Z} \exp\{\beta U(\pi_j)\}, \tag{12}$$

with a normalization coefficient

$$Z = \sum_j U(\pi_j) \exp\{\beta U(\pi_j)\}.$$

The parameter β characterizes the level of belief or confidence of the decision maker in the correct selection of the prospect set. Requiring that the utility factor, by its definition, be an increasing function of utility makes the belief parameter non-negative $(\beta \geq 0)$.

In the case of no confidence in the given set of prospects, we have

$$f(\pi_j) = \frac{U(\pi_j)}{\sum_j U(\pi_j)} \quad (\beta = 0). \tag{13}$$

In the opposite case of absolute confidence, we get

$$f(\pi_j) = \begin{cases} 1, & \pi_j = \max_j \pi_j \\ 0, & \pi_j < \max_j \pi_j \end{cases} \quad (\beta \to \infty), \tag{14}$$

where $\max_j \pi_j$ is the prospect whose expected utility is the largest. Thus, the latter situation $(\beta \to \infty)$ retrieves the deterministic utility theory, which hence can be seen as a particular case of the more general probabilistic approach.

The prospect utility factors $\{f(\pi_j) : j = 1, 2, \ldots, L\}$ give the fractions of decision makers selecting the corresponding prospects $\{\pi_j : j = 1, 2, \ldots, L\}$. The ordering of prospects in the probabilistic approach is the same as in the standard expected utility theory. But now, not all subjects are forced to choose the most useful prospect, though it is the prospect whose choice is the most probable. There can exist a fraction of decision makers choosing other prospects with lower utility. The probabilistic decision making scheme is summarized in Fig. 2.

Fig. 2 Scheme of the
probabilistic decision making
process based on the
evaluation of the prospect
utility factors characterizing
the fraction of decision
makers choosing the related
prospect

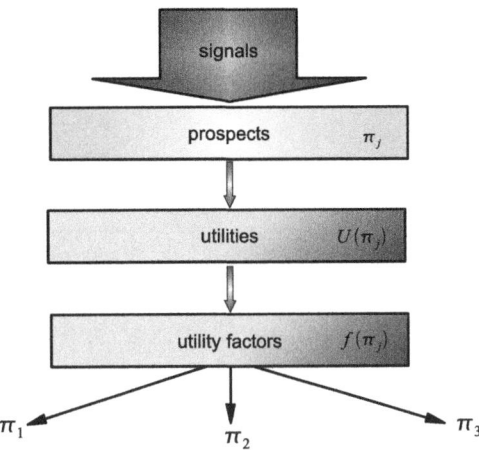

4 Human Decision Making and Computer Operations

It is widely believed that the human brain operates, during a decision making process, as a complex and powerful computer. The network of neurons within the brain accepts external signals and transforms them into decisions of the subject by accomplishing the corresponding actions [39]. Such a procedure could correspond to the schemes depicted in Figs. 1 or 2.

However, if the brain would act as just described, this would correspond to making decisions only on the basis of a well defined deterministic objective function, called utility. But there exist numerous empirical studies demonstrating that humans often deviate from and even contradict the choices prescribed by utility theory. Such contradictions are known as decision-making paradoxes. As examples, we can mention the Allais paradox [40], the Ellsberg paradox [41], the Kahneman-Tversky paradox [42], the Rabin paradox [43], the Ariely paradox [44], the disjunction effect [45], the conjunction fallacy [46], the planning paradox [47], and many others [48, 49]. These paradoxes cannot be resolved by the approaches consisting in modifying the expected utility theory into so-called non-expected utility theories, as has been proved in [50, 51].

The appearance of numerous paradoxes in decision making, based on utility theory, is caused by the fact that this theory does not take into account the emotional components always present in decision makers, which often compete and modify the decisions that would result purely from utility-based processes. A human decision maker not merely evaluates the objective utility of the prospects, but also is influenced by subjective feelings, emotions, and behavioral biases that are produced by subconscious brain activity. The brain takes decisions by combining (i) the objective knowledge of the prospect utility, by evaluating the utility factors, with (ii) the subjective attractiveness of the prospects, which is hinted by subconscious feelings. The latter means that, in addition to the utility factors measured by conscious logical operations, there should exist attraction factors produced by subconscious feelings. Then, the resulting weights of the prospects $p(\pi_j)$ are defined not merely by the

Fig. 3 Scheme of human
decision making, which is at
the basis of the Quantum
Decision Theory proposed
here

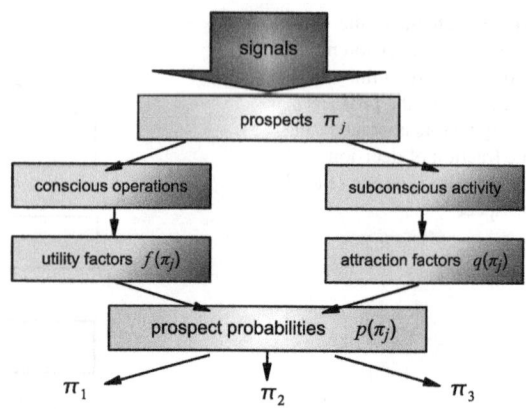

utility factors $f(\pi_j)$, but are also dependent on some attraction factors $q(\pi_j)$. We thus suggest that the correct representation of the brain function during a decision process is given by the scheme represented in Fig. 3, which should correct and replace those of Figs. 1 or 2.

Our theory views the human brain not just a powerful computer accomplishing a great number of straightforward logical operations, but as an object that must include parallel functioning on two levels. One part, representing conscious logical operations evaluating the utility factors, can be organized as a powerful computer. And the other part, representing subconscious activity producing the attraction factors, should be a very different device that functions not as a straightforward computer calculating numbers, but as an object estimating qualitative features of the prospects.

In the sequel, we do not touch on the technical issues of how the devices, discussed above, are actually structurally realized, or how they could be constructed in an artificial brain. Instead, we describe how their functioning can be represented mathematically, characterizing the split dual action of evaluating the prospect utilities and estimating their attractiveness.

5 Quantum Decision Making by Human Brain

The dualism of the brain, combining objective conscious operations with subjective subconscious activity, suggests that its functions could be described by generalizing the real-valued way of defining the prospect weights to an approach involving complex-valued quantities. In turn, this immediately points to quantum-theory techniques, where the probability weights are defined through complex-valued quantities, such as wave functions.

The idea of employing quantum theory for describing brain functions was advanced by Bohr [52, 53]. Analyzing the quantum theory of measurements, von Neumann [54] mentioned that the action of measuring observables could be interpreted, to some extent, as taking decisions. Using these ideas, we have developed [16–21] the Quantum Decision Theory (QDT), using the mathematical techniques of quantum theory of measurements.

Before formulating this theory, we would like to stress that the quantum approach to describing human decision making does not assume that the brain is a quantum object. The quantum techniques just provide the most straightforward way of generalizing decision making by taking into account the dual functioning of the human brain.

The main points of QDT are as follows. We consider a set of elementary prospects, represented by vectors $|n\rangle$, whose closed linear envelope

$$\mathcal{H} \equiv \mathrm{Span}\{|n\rangle\} \tag{15}$$

composes the space of mind. The prospects π_j from the given set $\mathcal{L}$ are represented by the vectors $|\pi_j\rangle$ in the space of mind. The prospect operators are defined as

$$\hat{P}(\pi_j) \equiv |\pi_j\rangle\langle\pi_j|. \tag{16}$$

These operators play the same role as the operators from the algebra of local observables in quantum theory.

The state of a decision maker is characterized by a non-negative operator $\hat{\rho}$ acting on the space of mind and normalized as

$$\mathrm{Tr}\,\hat{\rho} = 1,$$

with the trace taken over the space of mind. Defining the decision-maker state by a statistical operator, but not by a simple wave function, takes into account that this decision maker is not an absolutely isolated subject, but can be influenced by its environment.

The prospect probabilities, playing the role of observable quantities, are defined as the averages of the prospect operators

$$p(\pi_j) \equiv \mathrm{Tr}\,\hat{\rho}\,\hat{P}(\pi_j), \tag{17}$$

with the trace again taken over the space of mind. Writing down the explicit expression for the trace over the elementary prospect states and separating the diagonal and off-diagonal parts leads to the sum

$$p(\pi_j) = f(\pi_j) + q(\pi_j), \tag{18}$$

in which the first term comes from the diagonal part and the second term, from the off-diagonal part. The first term represents the classical utility factor, while the second term, caused by the prospect quantum interference, is the attraction factor. By definition, the prospect probability is non-negative and normalized, so that

$$\sum_{j=1}^{L} p(\pi_j) = 1, \quad 0 \leq p(\pi_j) \leq 1. \tag{19}$$

In view of the normalization condition for the utility factor (8), the attraction factor lies in the range

$$-1 \leq q(\pi_j) \leq 1 \tag{20}$$

and satisfies the *alternation law*

$$\sum_{j=1}^{L} q(\pi_j) = 0. \tag{21}$$

Generally, the attraction factor is a contextual quantity that can vary for different decision makers and even for the same decision maker at different times. This looks as an obstacle for the ability to give quantitative predictions for the prospect probabilities. However, it is possible to show [18, 21] that the aggregate attraction factor, averaged over many decision makers, enjoys the property called *quarter law*:

$$\frac{1}{L}\sum_{j=1}^{L} |q(\pi_j)| = \frac{1}{4}. \tag{22}$$

Since the utility factor is uniquely defined by the corresponding expected utility, it is possible to estimate *quantitatively* the prospect probabilities, assuming that the typical attraction factor satisfies the quarter law.

When the decision maker is a member of a society from which he/she gets additional information, then the attraction factor varies depending on the amount μ of the received information. The attraction factor, as a function of the information measure μ, can be presented [55] in the form

$$q(\pi_j, \mu) = q(\pi_j, 0)e^{-\gamma\mu}. \tag{23}$$

The information can be positive, with $\mu > 0$ as well as negative, or misleading, with $\mu < 0$. Respectively, the attraction factor can either decrease or increase. The attenuation of behavioral biases with the receipt of additional information has been confirmed by empirical studies [56, 57].

The reduction of QDT to the probabilistic variant of classical decision theory corresponds to the attraction factor tending to zero. This is similar to the reduction of quantum theory to classical statistical theory in the process of decoherence.

6 Cooperation Paradox in Prisoner Dilemma Games

Let us briefly summarize the status of QDT with respect to its empirical support. First, the disjunction effect, studied in different forms in a variety of experiments [45], has been analyzed in details in [18, 21], where we found that the empirically determined absolute value of the aggregate attraction factor $|q(\pi_j)|$ was found to coincide with the value 0.25 predicted by expression (22), within the typical statistical error of the order of 20 % characterizing these experiments. The same agreement, between the QDT prediction for the absolute value of the attraction factors and empirical values, holds for experiments testing the conjunction fallacy. The planning paradox has also found a natural explanation within QDT [17]. Moreover, it has been shown [20] that QDT explains practically all typical paradoxes of classical decision making, arising when decisions are taken by separate individuals.

In order to illustrate how QDT resolves classical paradoxes, let us consider a typical paradox happening in decision making. In game theory, there is a series of games, in which several subjects can choose either to cooperate with each other or to defect. Such setups have the general name of prisoner dilemma games. The cooperation paradox consists in the real behavior of game participants who often incline to cooperate despite the prescription of utility theory for defection. Below, we show that this paradox is easily resolved within QDT, which gives correct *quantitative predictions*.

The generic structure of the prisoner dilemma game is as follows. Two participants can either cooperate with each other or defect from cooperation. Let the cooperation action of one of them be denoted by C_1 and the defection by D_1. Similarly, the cooperation of the second subject is denoted by C_2 and the defection by D_2. Depending on their actions, the participants receive payoffs from the set

$$\mathbb{X} = \{x_1, x_2, x_3, x_4\}, \tag{24}$$

whose values are arranged according to the inequality

$$x_3 > x_1 > x_4 > x_2. \tag{25}$$

There are four admissible cases: both participants cooperate $(C_1 C_2)$, one cooperates and another defects $(C_1 D_2)$, the first defects but the second cooperates $(D_1 C_2)$, and both defect $(D_1 D_2)$. The payoffs to each of them, depending on their actions, are given according to the rule

$$\begin{bmatrix} C_1 C_2 & C_1 D_2 \\ D_1 C_2 & D_1 D_2 \end{bmatrix} \rightarrow \begin{bmatrix} x_1 x_1 & x_2 x_3 \\ x_3 x_2 & x_4 x_4 \end{bmatrix}. \tag{26}$$

As is clear, the enumeration of the participants is arbitrary, so that it is possible to analyze the actions of any of them.

Each subject has to decide what to do, to cooperate or to defect, when he/she is not aware about the choice of the opponent. Then, for each of the participants, there are two prospects, either to cooperate,

$$\pi_1 = C_1(C_2 + D_2), \tag{27}$$

or to defect,

$$\pi_2 = D_1(C_2 + D_2). \tag{28}$$

In the absence of any information on the action chosen by the opponent, the probability for each of these actions is $1/2$ (non-informative prior). Assuming for simplicity the linear utility as a utility function of the payoffs, the expected utility of cooperation for the first subject is

$$U(\pi_1) = \frac{1}{2}x_1 + \frac{1}{2}x_2, \tag{29}$$

while the utility of defection is

$$U(\pi_2) = \frac{1}{2}x_3 + \frac{1}{2}x_4. \tag{30}$$

The assumption of linear utility is not crucial, and can be removed by reinterpreting the payoff set (24) as the utility set. Because of condition (25), the utility of defection is always larger than that of cooperation, $U(\pi_2) > U(\pi_1)$. According to utility theory, this means that all subjects have always to prefer defection.

However, numerous empirical studies demonstrate that an essential fraction of participants choose to cooperate despite the prescription of utility theory. This contradiction between reality and the theoretical prescription constitutes the cooperation paradox [58, 59].

Considering the same game within the framework of QDT, we have the probabilities of the two prospects,

$$p(\pi_1) = f(\pi_1) + q(\pi_1), \qquad p(\pi_2) = f(\pi_2) + q(\pi_2). \tag{31}$$

The propensity to cooperation and the presumption of innocence propose that the attraction factor for cooperative prospect is larger than that for the defecting prospect, that is, $q(\pi_1) > q(\pi_2)$. In view of the alternation law (21) and quarter law (22), we have

$$q(\pi_1) = -q(\pi_2) = \frac{1}{4}. \tag{32}$$

Hence, we can estimate the considered prospects by the equations

$$p(\pi_1) = f(\pi_1) + 0.25, \qquad p(\pi_2) = f(\pi_2) - 0.25. \tag{33}$$

From here, we see that, even if defection seems to be more useful than cooperation, so that $f(\pi_2) > f(\pi_1)$, the cooperative prospect can be preferred by some of the participants.

To illustrate numerically how this paradox is resolved, let us take the data from the experimental realization of the prisoner dilemma game by Tversky and Shafir [45]. Subjects played a series of prisoner dilemma games, without feedback. Three types of setups were used: (i) when the subjects knew that the opponent had defected, (ii) when they knew that the opponent had cooperated, and (iii) when subjects did not know whether their opponent had cooperated or defected. The rate of cooperation was 3 % when subjects knew that the opponent had defected, and 16 % when they knew that the opponent had cooperated. However, when subjects did not know whether their opponent had cooperated or defected, the rate of cooperation was 37 %.

Treating the utility factors as classical probabilities, we have

$$f(\pi_1) = \frac{1}{2} f(C_1|C_2) + \frac{1}{2} f(C_1|D_2),$$
$$f(\pi_2) = \frac{1}{2} f(D_1|C_2) + \frac{1}{2} f(D_1|D_2).$$

According to the Tversky-Shafir data,

$$f(C_1|C_2) = 0.16, \qquad f(C_1|D_2) = 0.03.$$

Hence,

$$f(\pi_1) = 0.095, \qquad f(\pi_2) = 0.905. \tag{34}$$

Then, for the prospect probabilities (33), we get

$$p(\pi_1) = 0.345, \qquad p(\pi_2) = 0.655. \tag{35}$$

In this way, the fraction of subjects choosing cooperation is predicted to be about 35 %. This is in remarkable agreement with the empirical data of 37 % by Tversky and Shafir. Actually, within the statistical accuracy of the experiment, the predicted and empirical numbers are indistinguishable.

If we would follow the classical approach, the fraction of cooperators should be not larger than 10 % ($f(\pi_1)$), which is much smaller than the observed 37 %. But in QDT, there are no paradoxes and its predictions are in quantitative agreement with empirical observations.

7 Conclusion

We have presented the Quantum Decision Theory that we have developed since 2008, which is based on combining utility-like calculations with emotional influences in the representation of the decision making processes. We have emphasized that decision making by humans is principally different from the direct calculations by, even the most powerful, computers. This basic difference is in the duality of the human decision-making procedure. The brain makes decisions by a parallel processing of two different jobs: by consciously estimating the utility of the available prospects and by subconsciously evaluating their attractiveness.

We have shown how the duality of the brain functioning can be adequately represented by the techniques of quantum theory. The process of decision making has been described as mathematically similar to the procedure of quantum measurement. The self-consistent mathematical theory of human decision making that we have been developed contains no paradoxes typical of classical decision making. It is important to stress that this theory is the first theory allowing for it quantitative predictions taking into account behavioral biases.

We stress that the description of the functioning of the human brain by means of quantum techniques does not require that the brain be a quantum object, but this approach serves as an appropriate mathematical tool for characterizing the conscious-subconscious duality of the brain processes. This duality must be taken into account when one attempts to create an artificial intelligence imitating the human brain. Such an artificial intelligence has to be quantum in the sense explained above [60].

Acknowledgements The authors are grateful for many discussions with and advice from M. Favre and E.P. Yukalova. Financial support form the Swiss National Science Foundation is appreciated.

References

1. F.A.C. Azevedo, L.R.B. Carvalho, L.T. Grinberg, J.M. Farfel, R.E.L. Ferretti, R.E.P. Leite, W.J. Filho, R. Lent, S. Herculano-Houzel, Equal numbers of neuronal and non-neuronal cells

make the human brain an isometrically scaled-up primate brain. J. Comp. Neurol. **513**, 532–541 (2009)

2. G.M. Shepherd, *Neurobiology* (Oxford University, Oxford, 1994)
3. E.R. Kandel, J.H. Schwartz, T.M. Jessel, *Principles of Neural Science* (McGraw-Hill, New York, 2000)
4. O. Sporns, *Networks of the Brain* (Massachusetts Institute of Technology, New York, 2011)
5. A.R. Luria, *Higher Cortical Functions in Man* (Basic Books, New York, 1966)
6. D. Elkind, J. Flavell, *Studies in Cognitive Development* (Oxford University, New York, 1969)
7. R.J. Sternberg, W. Salter, *Handbook of Human Intelligence* (Cambridge University, Cambridge, 1982)
8. J.P. Das, J.A. Naglieri, J.R. Kirby, *Assessment of Cognitive Processes* (Allyn and Bacon, Needham Heights, 1994)
9. W.K. Wake, H. Gardner, M.L. Kornhaber, *Intelligence: Multiple Perspectives* (Harcourt Brace College, Fort Worth, 1996)
10. K. Richardson, *The Making of Intelligence* (Columbia University, New York, 2000)
11. R.J. Sternberg (ed.), *International Handbook of Intelligence* (Cambridge University, Cambridge, 2004)
12. K. Stanovich, *What Intelligence Tests Miss: The Psychology of Rational Thought* (Yale University, New Haven, 2009)
13. S. Coren, *The Intelligence of Dogs* (Bantam Books, New York, 1995)
14. A. Trewavas, Green plants as intelligent organisms. Trends Plant Sci. **10**, 413–419 (2005)
15. J. Canny, S.J. Russell, P. Norvig, *Artificial Intelligence: A Modern Approach* (Prentice Hall, Englewood Cliffs, 2003)
16. V.I. Yukalov, D. Sornette, Quantum decision theory as quantum theory of measurement. Phys. Lett. A **372**, 6867–6871 (2008)
17. V.I. Yukalov, D. Sornette, Physics of risk and uncertainty in quantum decision making. Eur. Phys. J. B **71**, 533–548 (2009)
18. V.I. Yukalov, D. Sornette, Processing information in quantum decision theory. Entropy **11**, 1073–1120 (2009)
19. V.I. Yukalov, D. Sornette, Entanglement production in quantum decision making. Phys. At. Nucl. **73**, 559–562 (2010)
20. V.I. Yukalov, D. Sornette, Mathematical structure of quantum decision theory. Adv. Complex Syst. **13**, 659–698 (2010)
21. V.I. Yukalov, D. Sornette, Decision theory with prospect interference and entanglement. Theory Decis. **70**, 283–328 (2011)
22. J. von Neumann, O. Morgenstern, *Theory of Games and Economic Behavior* (Princeton University, Princeton, 1953)
23. L.J. Savage, *The Foundations of Statistics* (Wiley, New York, 1954)
24. J.W. Pratt, Risk aversion in the small and in the large. Econometrica **32**, 122–136 (1964)
25. K.J. Arrow, *Essays in the Theory of Risk Bearing* (Markham, Chicago, 1971)
26. D. Graeber, *Debt: The First 5,000 Years* (Melville House, New York, 2011)
27. D. Bernoulli, Exposition of a new theory on the measurement of risk. Proc. Imper. Acad. Sci. St. Petersbg. **5**, 175–192 (1738)
28. D. McFadden, Econometric models of probabilistic choice, in *Structural Analysis of Discrete Data with Econometric Applications*, ed. by C.F. Manski, D. McFadden (Massachusetts Institute of Technology, Cambridge, 1981), pp. 198–272
29. I.M. Rutkow, Surgical decision making, the reproducibility of clinical judgment. Arch. Surg. **117**, 337–340 (1982)
30. J.O. Nielsen, H. Dons-Jensen, H.T. Sarrensen, Lauge-Hansen classification of malleolar fractures, an assessment of the reproducibility in 118 cases. Acra Orthop. Scand. **61**, 385–387 (1990)
31. J.W. Gibbs, *Elementary Principles in Statistical Mechanics* (Oxford University, Oxford, 1902)
32. J.W. Gibbs, *Collected Works*, vol. 1 (Longmans, New York, 1928)

33. J.W. Gibbs, *Collected Works*, vol. 2 (Longmans, New York, 1931)
34. C.E. Shannon, W. Weaver, *Mathematical Theory of Communication* (University of Illinois, Urban, 1949)
35. E.T. Jaynes, Information theory and statistical mechanics. Phys. Rev. **106**, 620–630 (1957)
36. V.I. Yukalov, Phase transitions and heterophase fluctuations. Phys. Rep. **208**, 395–492 (1991)
37. S. Kullback, R.A. Leibler, On information and sufficiency. Ann. Math. Stat. **22**, 79–86 (1951)
38. S. Kullback, *Information Theory and Statistics* (Wiley, New York, 1959)
39. K. Mainzer, *Thinking in Complexity* (Springer, Berlin, 2007)
40. M. Allais, Le comportement de l'homme rationnel devant le risque: critique des postulats et axiomes de l'ecole Americaine. Econometrica **21**, 503–546 (1953)
41. D. Ellsberg, Risk, ambiguity, and the Savage axioms. Q. J. Econ. **75**, 643–669 (1961)
42. D. Kahneman, A. Tversky, Prospect theory: an analysis of decision under risk. Econometrica **47**, 263–291 (1979)
43. M. Rabin, Risk aversion and expected-utility theory: a calibration theorem. Econometrica **68**, 1281–1292 (2000)
44. D. Ariely, *Predictably Irrational* (Harper, New York, 2008)
45. A. Tversky, E. Shafir, The disjunction effect in choice under uncertainty. Psychol. Sci. **3**, 305–309 (1992)
46. A. Tversky, D. Kahneman, Extensional versus intuitive reasoning: the conjunction fallacy in probability judgement. Psychol. Rev. **90**, 293–315 (1983)
47. F.E. Kydland, E.C. Prescott, Rules rather than discretion: the inconsistency of optimal plans. J. Polit. Econ. **85**, 473–492 (1977)
48. C.F. Camerer, G. Loewenstein, R. Rabin (eds.), *Advances in Behavioral Economics* (Princeton University, Princeton, 2003)
49. M.J. Machina, Non-expected utility theory, in *New Palgrave Dictionary of Economics*, ed. by S.N. Durlauf, L.E. Blume (Macmillan, New York, 2008)
50. Z. Safra, U. Segal, Calibration results for non-expected utility theories. Econometrica **76**, 1143–1166 (2008)
51. N.I. Al-Najjar, J. Weinstein, The ambiguity aversion literature: a critical assessment. Econ. Philos. **25**, 249–284 (2009)
52. N. Bohr, Light and life. Nature **131**, 421–423 (1933), 457–459
53. N. Bohr, *Atomic Physics and Human Knowledge* (Wiley, New York, 1958)
54. J. von Neumann, *Mathematical Foundations of Quantum Mechanics* (Princeton University, Princeton, 1955)
55. V.I. Yukalov, D. Sornette, Role of information in decision making of social agents. Int. J. Inf. Technol. Decis. Mak. (in press)
56. A. Kühberger, D. Komunska, J. Perner, The disjunction effect: does it exist for two-step gambles? Organ. Behav. Hum. Decis. Process. **85**, 250–264 (2001)
57. G. Charness, E. Karni, D. Levin, On the conjunction fallacy in probability judgement: new experimental evidence regarding Linda. Games Econ. Behav. **68**, 551–556 (2010)
58. C. Camerer, *Behavioral Game Theory* (Princeton University, Princeton, 2003)
59. A. Tversky, *Preference, Belief, and Similarity: Selected Writings* (Massachusetts Institute of Technology, Cambridge, 2004)
60. V.I. Yukalov, D. Sornette, Scheme of thinking quantum systems. Laser Phys. Lett. **6**, 833–839 (2009)

From Neurons to Neutrinos—Modeling the Whole Earth System

Robert Bishop

1 Introduction

This paper was written in conjunction with a talk given as part of a symposium held in honor of Dr. Minoru Freund, friend of the author and highly respected scientist at the NASA AMES Research Center, Mountain View, California.

Mino was indeed a global thinker and a Renaissance man—he was able to tackle the physics of space as well as the horrendous complexity of his own brain tumor and many levels of science in between. He understood that *Modeling the Whole Earth System* is a challenge whose time has come. Mino knew that we now have the tools for the job, and because of this he encouraged me to proceed with all speed. The title of my talk and of this paper reflects his encouragement.

2 The Nature of the Challenge

To contemplate the *Modeling the Whole Earth System* challenge, one must first think of its multidimensionality, simply because the many facets of the Earth itself—from solid core, plastic mantle, brittle crust, shifting continents, deep oceans, and thin atmospheric skin, out to the remote but powerful influences of our Sun and Moon, and further out still to the distant heavenly bodies within our Solar System—are all part of a dynamic interplay. And our level of concern does not stop there, since incoming cosmic rays and high-energy particles, which impact Earth's upper atmosphere, emanate from sources well beyond our Solar System.

For all such physical aspects of the cosmos, we are especially fortunate to live at a time when massive amounts of data are being collected and collated on a daily

R. Bishop (✉)
Founder and President, ICES Foundation, Geneva, Switzerland
url: http://www.icesfoundation.org
url: http://www.youtube.com/watch?v=WN1bI8RgRIY&feature=youtu.be

F. Freund, S. Langhoff (eds.), *Universe of Scales: From Nanotechnology to Cosmology*,
Springer Proceedings in Physics 150, DOI 10.1007/978-3-319-02207-9_12,
© Springer International Publishing Switzerland 2014

basis—indeed; much of this effort is due to the extensive capabilities of NASA and its global partners and programs.

This summer for instance, we witnessed the 40-year anniversary of the highly successful *Landsat* program, the retirement of the Space Shuttle and its major contributions to Earth observation, and NASA's brilliant *Curiosity* rover landing on the surface of Mars—each one of these occasions being testimony to the remarkable technologies at hand.

Yet the most difficult aspect of the *Modeling* challenge lies right at our doorstep—namely, a clear understanding of how the many varieties of life that abound upon the planet integrate and impact on the underlying physics and chemistry of their rocky domain. Can we ever comprehend the full complexity of life and its impact on the planet itself? Is it possible to mathematically abstract the essential elements of biology? Are social systems within reach of our equations and models? And is it possible to integrate and model *the whole system*—natural sciences and socioeconomic sciences, interacting as one?

Such questions will certainly cause serious debate, yet discoveries in the 21st Century will create many surprises, possibly resolving these puzzles, so we should not be deterred from launching our mission and pursuing the very ambitious goal of *Modeling the Whole Earth System* at this point in time.

3 The Blue Marble

Almost 40 years ago, on 7 December 1972 at 10:39 UTC from 45,000 kilometers above Earth onboard Apollo 17, Dr. Harrison Schmitt, the only geologist to have flown in space, took one of the most famous photographs of the space era. Dubbed by NASA as the *Blue Marble* and shown in Fig. 1, this was not a digital picture, but a picture taken on regular chemical film with a large format Hasselblad camera.

With Antarctica centered at the bottom of the frame surrounded by large scale storms circling above the Southern Ocean, the view extends northwards to the African land mass, Madagascar and the Arabian Peninsula—with the Indian and Atlantic oceans on either side. Most likely this picture changed the worldview of human society at that time, for intuitively, we could finally understand that our planet was fragile and something to be protected. It showed the Earth as an integrated and borderless whole.

Yet in reality at ground level we have over 200 nation states dividing the world politically, and in the sciences and humanities we have several hundred siloed and stovepipe specializations that divide our thinking and belief system. The challenge at hand calls for a significant *re-integration* of these matters.

In a practical sense however, we need the methods, tools and data that can move us in this direction, along with a number of simple strategic insights. This paper will attempt to bring together such items, pointing out how progress in such matters over the last few decades has been stop and go. Given the string of disasters that has beset the planet during the same period however, and fearing that more is to come, we would be all well advised to shift our reintegration efforts into a higher gear.

Fig. 1 The Blue
Marble—taken from
45,000 km on December 7,
1972 at 10:39
UTC—courtesy NASA

4 Vast Pools of Data Are Now at Our Fingertips!

Observation instruments located on Earth's surface as well as in space are now generating petabytes of data per day, and by mid century will be generating exabytes per day, possibly zettabytes and ultimately yottabytes of data per day.[1] This is indeed big data and it is progressively delivered in rich, streaming, high-resolution digital format, more than doubling in quantity each year. Furthermore, the efforts and instruments of *citizen scientists* are providing an increasing percentage of this data.

Socioeconomic data is headed along the same growth path—with personal records and public services now having been transformed into digital format in most all aspects of modern life: government, financial, legal, medical, educational, and especially in entertainment media such as film, TV, radio, social net-working, blogs and ebooks—frequently with the aid of inexpensive digital video cameras.

The low cost and ubiquity of pdas, cellphones, iPads and laptops, together with their cameras and service networks, has ensured that both natural sciences and socioeconomic sciences can beneficially collect and process data at will. In addition, using miniaturized sensors, many products and processes now embody *local intelligence*, acting more autonomously than they could have in the past.

Except in rare cases, we can say our main problem is no longer the collection of data sets, but how to move them, store them, harmonize and curate them, and then ultimately *federate* them, so that others can access them, verifying before assimilating them into various computational models. Integrative computational models can frequently clarify data sets whose sheer size would otherwise place them outside of human ken.

---

[1]Petabytes are 10^{15} bytes, exabytes are 10^{18} bytes, zettabytes are 10^{21} bytes, and yottabytes are 10^{24} bytes.

5 The Role of IT Power Tools

Dealing with this *exaflood* of data is the role of IT power tools—turning data into information, information into theory, and hopefully, theory into knowledge and even wisdom. These tools comprise a set of processes and procedures well known to scientists today and progressively evaluated by the public at large, namely: data mining and analytics, modeling and simulation, and interactive immersive visualization.

Depending on the complexity of the work at hand, such power tools can be brought to bear on big data by means of high performance computing platforms, cloud computing, grid computing or a network of regular desktop PCs. Truthfully, it can be said that we'll not fully understand any natural phenomenon, whether it be physical or biological, until it can be computationally modeled, simulated and visualized.

Nonetheless, our power tools and data sets have not served their ultimate purpose until having helped us uncover the hidden and holistic insights which underpin this dynamic planet of ours—insights that not only educate us, but insights that help us protect the planet itself; insights that encourage society to grow in a harmonious way; insights that preserve biodiversity and keep people safe.

One could clearly argue from the frequency of global disasters that we must urgently make societies more resilient as well. Indeed, the making of global policy will need serious enhancement if we are to govern the planet in a more coherent and balanced manner. Evidence-based policies are needed to create an equitable distribution of opportunity for all nations, and for all levels of society therein.

These are the higher-level goals that future policy making can help achieve—bringing natural and socioeconomic sciences together and deploying the most advanced technologies, methods and processes available. Not a very easy task, but achievable in the decades ahead if pursued steadfastly.

6 How Bad Can It Get?

The month of July 2012 was the hottest on US record wherein 40,000 heat records were already established by mid-year. Additionally, in what became an unrelenting heatwave, 63 % of lower 48 states were declared to be in drought, the worst to hit the US since 1956.[2]

Other forms of extreme weather are also occurring across the USA weekly, including wild-fires, floods, violent storms, derechos,[3] tornadoes and hurricanes, all of which wreak havoc when they strike—yet it is still unclear if these extreme events can be attributed to global warming, or not.

---

[2]See BofA Merrill Lynch Global Research Report 'Global Drought—Opportunities and Risks'.

[3]Derechos are long-lived straight-line windstorms associated with a fast band of thunderstorms.

If one mixes human design mistakes with Nature's fury however, then damage can multiply exponentially—as happened in August 2005 when hurricane Katrina hit the New Orleans levy system. It also happened with the Deepwater Horizon oil spill of April-July 2010, with almost 5 million barrels from this spill devastating the economy and coastline of southern US Gulf states.

As for circumstances outside of North America, things could hardly be worse—numerous super-typhoons have hit the Philippines, Taiwan, China, Japan and Korea, causing thousands of lives to be lost, as have extra-ordinary monsoonal floods in India, Pakistan, Bangladesh, Myanmar, Thailand, Cambodia, Laos and Vietnam. In 2011, Australia witnessed a direct hit by Category 5 cyclone Yasi, and monsoonal Queensland floods commencing in the previous year were of biblical proportions.

Perhaps what was previously called the *storm of the century* has become the *storm of the decade*. Or perhaps we are simply witnessing multiple *black swan* events that are coincidental in their occurrence?

In contrast to rain and storms, Russia experienced a devastating heatwave in 2010 similar to that of France in 2003—thousands died in both cases. And volcanic ash clouds have closed down airspace in recent years, affecting all of Europe in one instance, and Russia and Latin America at other times—causing thousands of flight cancellations and millions of passenger rebookings in response.

Yet all of this pales in comparison to the devastating earthquakes that have occurred around the planet during this last decade, often times causing large tsunamis. For example, the December 2004 Banda Aceh M9.2 earthquake and tsunami taking over 200,000 lives; the May 2008 earthquake killing nearly 70,000 people in Szechuan, China; the Haitian earthquake of January 2010 with thousands dead; and the multiple earthquakes in Iran, Turkey, Italy, Chile and New Zealand, all with heavy fatalities.

This brings us to Tohoku Japan, where in March 2011 *multiple synchronous collapse* occurred—the cascading effect of risks transferred from one interconnected area of society to another. No event illustrates this domino-like phenomenon more than the Great East Japan Earthquake, the corresponding tsunami, shown in Fig. 2, and the subsequent melt down of three nuclear reactors at the Fukushima Daiichi plant.

In this event, nearly 400,000 homes were swept away, 19,850 people lost their lives, several villages were evacuated because of contamination, regional food and water supplies were rendered useless, the Tokyo Electric Power Company (TEPCO) was nationalized, the Prime Minister of Japan was replaced, and nuclear safety standards around the world were forcibly upgraded. Furthermore, as a result of this disaster, Germany and Switzerland decommitted from their nuclear power strategies, and to make matters worse, a 2012 scientific report[4] has since indicated that there are presently 23 nuclear sites with a total of 74 nuclear reactors sitting in tsunami vulnerable areas of the world today.

---

[4]Natural Hazards 63(2), 1273–1278, doi:10.1007/s11069-012-0162-0, September 2012.

Fig. 2 Multiple synchronous collapse—the tsunami of March 2011 arriving at the shores of To-hoku region, Japan

7 We Have Built a *Brittle* Social Fabric

From the combined impact of the quarterly profit demands of Wall Street, the minimal reserves of *just-in-time* manufacturing, and the *stovepiping* of knowledge and operations into countless specialized departments of government, academia and industry, we have indeed built a complex and brittle modern society containing a vast number of non-linear feedback loops. This should be seen as *optimized complexity*—a society with low planned reserves and thin safety margins—where few people can see the forest from the trees, and fewer still can detect the precursor signals that are provided by circumstances before disaster strikes. The scale of the problem at hand is exemplified in Fig. 3.

Black swan events are by definition a surprise and presuppose that there are no precursor signals, which in fact is *not* the case. A society with deep cross-connections within its knowledge base will see and read the right signals at the right time. Re-integration is a solution to this conundrum—an answer to present day problems of an advanced society with over specialization of its people, its knowledge systems and its know-how. If we only knew what in fact we already know!

This is the main reason I believe that our planet has come very close to transgressing a number of *planetary boundaries*—safe operational limits—whereby, beyond those limits, runaway conditions apply, and irreversible dynamics take us into the danger zone. We are not sharing what we know.

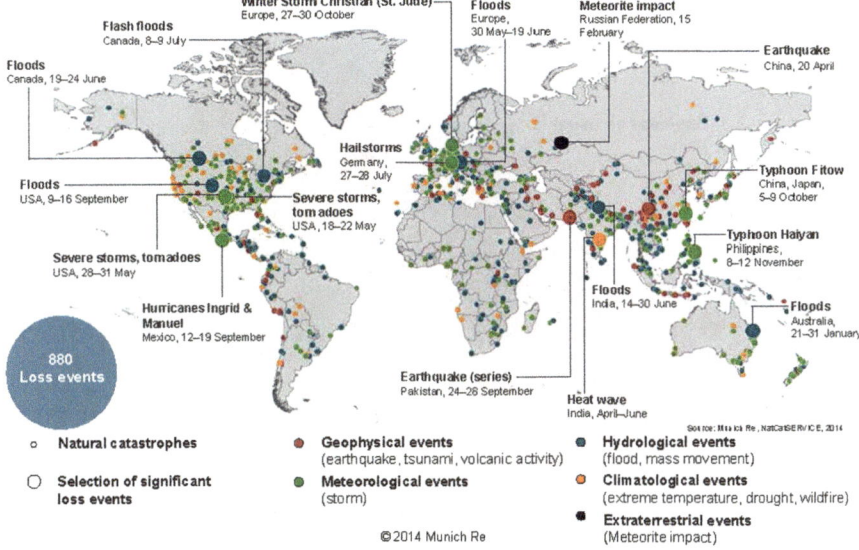

Fig. 3 World map of natural catastrophes for January to June 2012—courtesy Munich Re

8 Planetary Boundaries

Through the work of the Stockholm Resilience Centre,[5] nine *boundary rails* have been defined and quantified, a number of which we are already in violation of. Perhaps we can successfully retreat from such a precarious position, perhaps not.

The nine rails in contention pertain to atmospheric CO_2, stratospheric ozone depletion, aerosol loading of the atmosphere, global phosphorous and nitrogen cycles, ocean acidification, freshwater usage, land use changes, biodiversity loss and chemical pollution of the environment.

It is clear that if we violate too many of these boundaries simultaneously, there becomes a high risk that the planet's immune system will respond with its own form of self-defense—*a mass extinction event*. Such an event appears to have happened on Earth several times before in its remote past.

Figure 4 below graphically illustrates this scenario, whereas in Fig. 5a the mathematics of one single variable *lambda* is shown, the net radiation damping coefficient, as it swings the state of the planet from a safe equilibrium condition to runaway global warming behavior.

Figure 5b illustrates how critical the current state of affairs has become, by comparing the estimated *climate sensitivity parameter*, as calculated by each of several models. In the model labeled 'Earth System' whereby all identified feedback loops are included in the calculation, should the concentration of CO_2 parts per million in the atmosphere reach a level of 560 ppm (double that at the beginning of

---

[5]Part of the Stockholm University: http://www.stockholmresilience.org/planetary-boundaries.

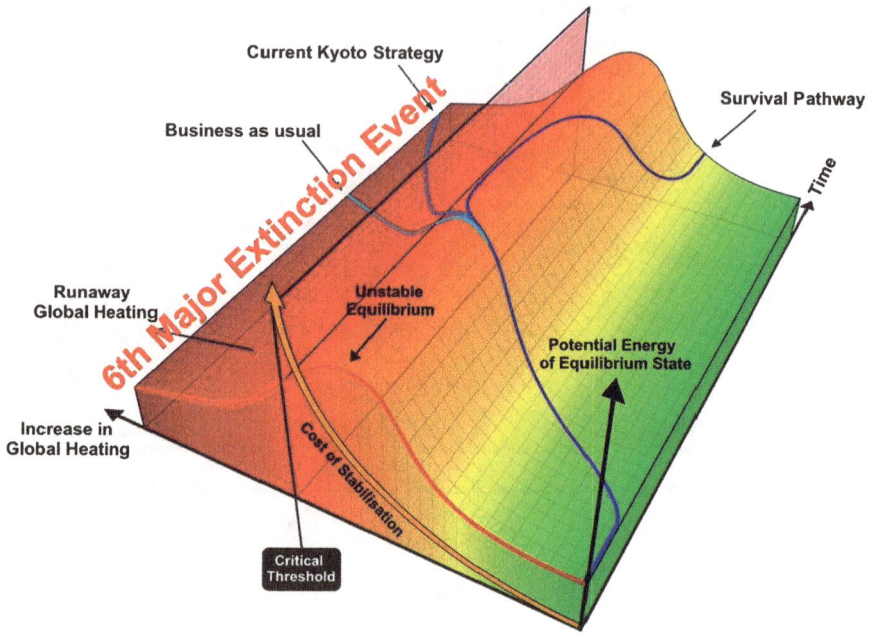

Fig. 4 Pictorial illustration of global warming exceeding critical threshold, or retreating to survival pathway

the Industrial Revolution), then the model predicts that average global surface-level temperatures will climb by 7.8 degrees Celsius, far above current expectations, and above each of the other four models shown on the graph. Because this latter model includes most of the known feedback loops operating within the Earth System to-day (Fig. 6), it can be considered as more complete than the other four comparative models shown. Reproduction of these charts is by permission of the author.[6]

9 The UN Call to Action

Over the years there have been many alerts issued by the United Nations and its agencies for world leaders and nations to come together and agree upon new in-dustrial, agricultural and social behavior aimed at capping the ever rising amount of carbon dioxide in the atmosphere, if not diminishing it. Unfortunately, there has also been intense debate during this period, firstly as to which countries are to blame, and secondly as to the root cause of global warming—anthropogenic versus natural variability. Although the scientific community has solidly implicated anthropogenic

[6]David Wasdell, Director of the Apollo-Gaia Project, London, UK. Full analysis is available at http://www.apollo-gaia.org/Climate_Sensitivity.htm.

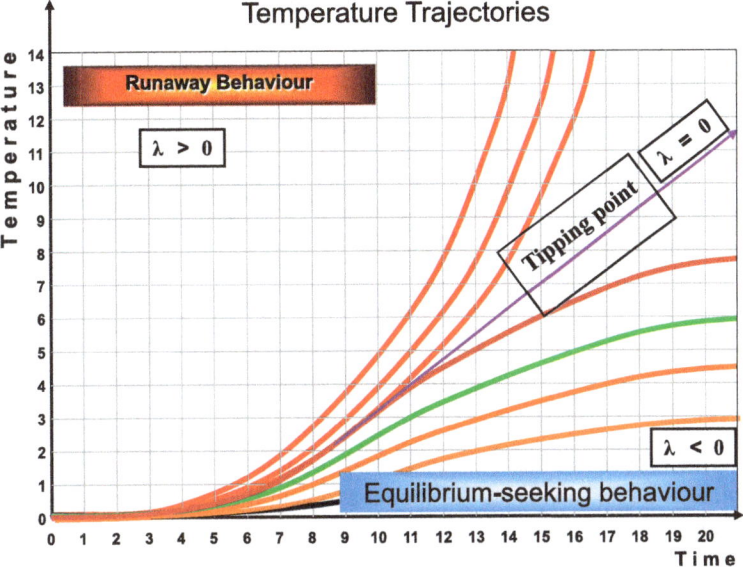

Fig. 5a Thermodynamic radiation model where lambda = net radiation damping coefficient

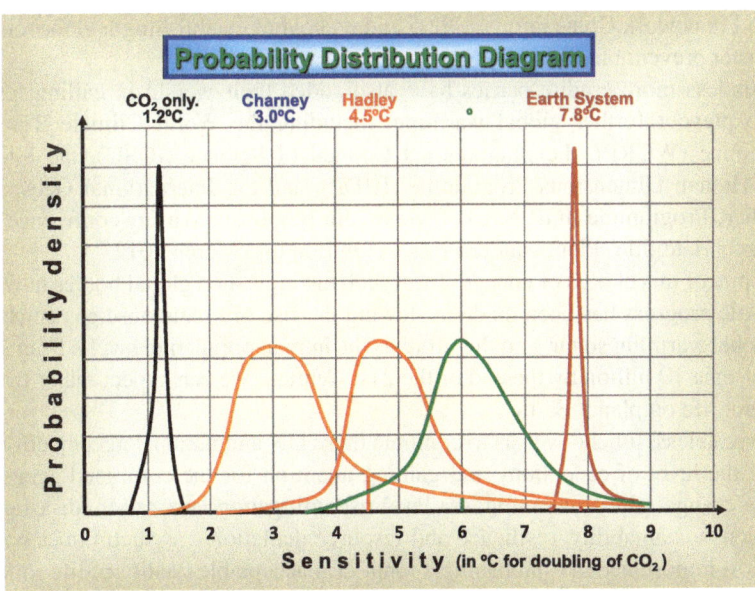

Fig. 5b Climate Sensitivity showing change in degrees C if CO_2 is doubled from 280 ppm to 560 ppm

Fig. 6 Multiple non-linear feedback loops in the total Earth System—some positive, some negative

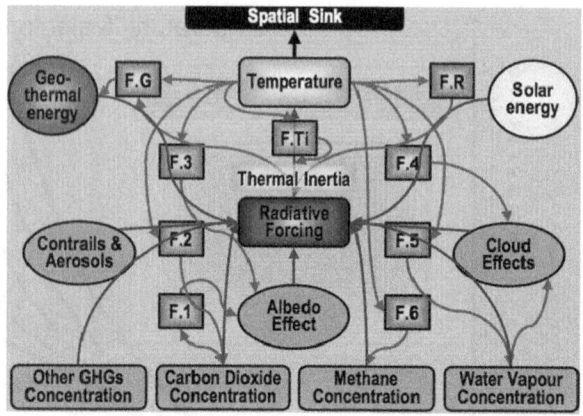

factors as the root cause, this evidence-based opinion has yet to be embraced by the body politic at large, and so nations continue to pour their GHG[7] emissions into the planet's atmosphere at an alarming rate, and without remorse.

Thought leadership on these issues has been provided by the Intergovernmental Panel on Climate Change (IPCC) established by the UN in 1988; the UN Framework Convention on Climate Change (UNFCCC) established in 1992 with its Kyoto Protocol of 1997 and recent follow-up conference of the parties in Copenhagen, Cancun and Durban; the UN Development Programme (UNDP) with its Millennium Development Goals and Rio Earth Summits of 1982 and 2012; and the UN International Strategy for Disaster Reduction (UNISDR) established in 1999 with its Hyogo Framework Convention of 2005 and its goal of establishing a global culture of disaster prevention.

Countless more erudite parties have also added their weight in calling for action to prevent further global warming, including the World Climate Research Programme (WCRP), the International Council of Science (ICSU), the International Human Dimensions Programme (IHDP), and the International Geosphere-Biosphere Programme (IGBP), all of whom jointly sponsored a key conference with the appropriate title 'Planet under Pressure' in London, March 2012.

In spite of this chorus of concern from such distinguished global bodies however, very little progress has been made in slowing the rise of greenhouse gas emissions and global warming so far, and therefore as the human population swells from 7 billion to some 10 billion by the end of the 21st Century, we can expect major trouble ahead for life on planet Earth.

Nevertheless, it behooves us to continue these UN and national agency efforts—raising the level of cacophony and gaining attention for the combined agenda of climate change, global warming, sea level rise, mitigation, adaptation, disaster risk reduction, sustainability, resilience and resource depletion—as such issues will be of utmost importance in establishing a long-term acceptable quality of life for all.

---

[7]Green House Gas.

In case we are unsuccessful in these efforts however, there is an emerging Plan B, but it too threatens the planet in its own way. Plan B calls for humans to experiment with *geoengineering*—taking risks to alter irradiance received at ground level from the Sun for example, or changing the albedo of planetary landscapes and oceans. These ideas are under investigation at several science centers around the globe, but needless to say, all such concepts have possible unintended consequences, and it is in predicting these long-term effects that our current knowledge and models are insufficient.

Of course, ten thousand years of civilization and the insatiable demands for food, water, clothing, housing, transportation and energy have already caused substantial geoengineering of planet Earth by human society. Plan B calls for a continuation of this process but in a much more accelerated manner. It is an additional reason for society to hasten the build up of modeling and simulation skills, so as to fully understand the *what ifs* of certain Plan B actions, should they require to be taken in the years ahead.

10 Coupled Model Intercomparison Project (CMIP)

The US Department of Energy's Program for Climate Model Diagnosis and Intercomparison (PCMDI), along with WCRP and IGBP, are coordinating Phase 5 of this important project and aiming to compare the *skill* of 61 climate models being submitted from 27 modeling centers around the world. Data will be held in a distributed archive led by the Earth System Grid Federation (ESGF), and results will become part of the IPCC Assessment Report 5, due in late 2013.

CMIP5 promotes a standard set of model simulations in order to *hindcast* the recent past; *forecast* the future climate at years 2035 and 2100, respectively, and quantify key feedbacks involving clouds and the global carbon cycle. No doubt, there will be considerable dispersion of results among the 61 climate models, and so elaborate statistical methods will need to be deployed to arrive at a useful conclusion.

In addition, there is the open question of coupling climate models to the environment itself and to society at large, including how to introduce feedback loops that connect all three aspects of the whole Earth System. This is where Integrated Assessment Modeling (IAM) becomes necessary.

IAM allows various social response scenarios to force climate model outputs—introducing the possible mitigation and adaptation actions of society, or changes in demographics, technology usage and future levels of economic activity. Such *social development pathways* play an important role in determining the long-term behavior of the entire Earth System, yet they have only played a very minor role in earlier phases of CMIP analyses so far.

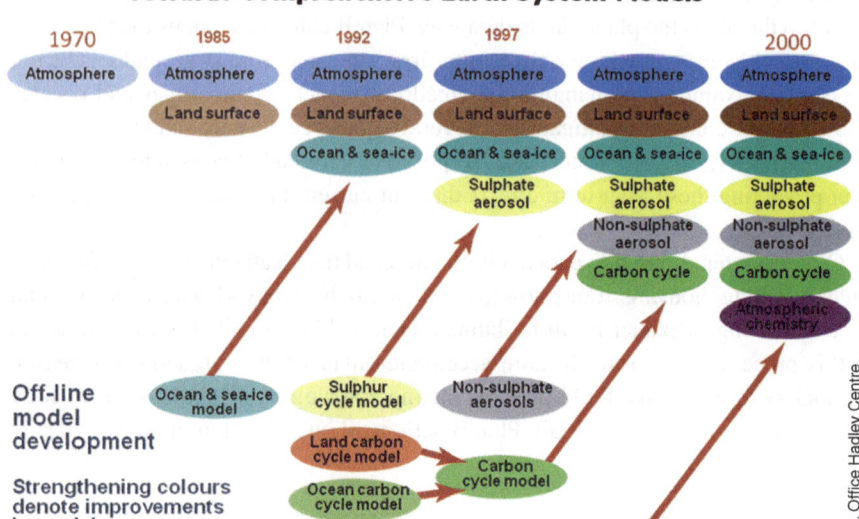

Fig. 7 Thirty years of progressive model development showing incremental components being accreted

11 Towards a Globally-Focused Earth System Simulator

We might ask then: can we ever fully integrate our models so as to view the Earth System both globally and holistically? Or are our social, educational, governmental, national and international attitudes too specialized and inflexible to embrace the whole? Can any one country or any one organization supply the full portfolio of knowledge to do so, and can any one entity provide the underlying compute capacity to achieve the right results? Is it even affordable in this era of national budget constraints?

Mathematically, it is a very big challenge to link and couple models from disparate fields—each deploying its own branch of algorithmic formulation, and having different grid-point systems and time-stepping techniques. Yet within the reality of Nature itself we find a smooth cross-linking of multiple domains, so we should at least be confident that it can be done. Figure 7 shows the steady progress that has been made over a 30-year period within the climate modeling community in building comprehensive Earth System models. This is progress indeed, however, the job will not be complete until we can fully couple weather, climate, environment and ecosystems with models of the entire biosphere, and couple this accretion in turn with models of the Earth's tectonic, crust, mantle and core. From there we need to further integrate models of the Earth's ionosphere and magnetosphere, and in doing so ultimately embrace the interactions of the Earth System with incoming space weather and the continuous stream of coronal mass ejections (CMEs) emanating from our Sun. And let's not forget the powerful gravitational effects of both the Sun and the Moon, which not only generate our ocean tides, but also dictate the reproduction cycle of every living species on the planet.

Taking into account the enormous scope of this challenge, most logically it should be pursued as an international effort, for example along the lines of CERN—the European Organization for Nuclear Research—located astride the Franco-Swiss border near Geneva. CERN has been in existence since 1954 and is financially supported by 20 Member States. Other organizational bodies of a scientific nature include the European Centre for Medium-Range Weather Forecasts in Reading, UK (ECMWF), and the International Thermonuclear Experimental Reactor in Cadarache, France (ITER).

In our case however, time is of the essence and there are very few government funds available for immediate action on the scale of prior efforts. Nonetheless, we can lay down the philosophical foundations for moving forward and outline the appropriate set of guiding principles to follow:

- a not-for-profit, politically independent, agile, fast moving public-private-partnership
- an organization with philanthropic funding support and fee-based consulting revenue
- committed to open science, open data files, open source code & open access publishing
- strong scientific participation, socioeconomic reach, and policy guidance capabilities
- networked worldwide into government organizations, non-government organizations, universities, research organizations, private industry and citizen science activities.

In January 2010, a Swiss entity under the name of the *ICES Foundation* was established with the above principles in mind—see Fig. 8. This entity is now preparing to take on the challenge of *Modeling the Whole Earth System*, and is actively pursuing appropriate global scientific partnerships and philatropic funding. The ICES Foundation will actively seek to partner with US entities: NASA, NOAA, LLNL, PNNL, ORNL, NCAR, GFDL, IGES, USGS, and CALIT2, and with their counterparts in other countries. To enhance global efforts, the ICES Foundation will devote 25 % of its computing resources in assisting developing countries which have limited resources of their own, or no resources at all.

12 The Future of Computing Platforms

It is not the intention of the ICES Foundation to collect Earth Observations of its own accord, but simply to access via established portals the data that has been collected by others. GEO[8] and GOSIC[9] are two such portals. Likewise, it is not the

---

[8]GEO: Group on Earth Observations http://www.earthobservations.org/index.shtml.

[9]GOSIC: Global Observing Systems Information Center: http://gosic.org/.

Fig. 8 The ICES Montage
was created by Tony and
Bonnie DeVarco for the ICES
Foundation with use of
satellite imagery courtesy of
NASA's Earth Observatory.
The inset diagram is a
simplified version of the *Tree
of Life*, showing common
names of the major groups.
This version of the tree is
based on the *Tree of Life
appendix in Life: The Science
of Biology*, 8th ed., by
D. Sadava, H.C. Heller,
G.H. Orians, W.K. Purves,
and D.M. Hillis (Sinauer
Associates and
W.H. Freeman, 2008). Other
insets include the Sun Dagger
and the Analemma. The
constellation Pleiades peeks
through in the background.
© ICES Foundation

intention of the ICES Foundation to rewrite any of the computing models that have already been written, and which are available to the research community. Such models include CESM[10] and the multiple models within IS-ENES.[11] However, coupling strategies within these models as well as adding more elements to the models *is* the mission of ICES—and so is the integration of newly emerging scientific knowledge into such models.

Because of the rapid transformation to massive parallelism in the architecture of present day compute platforms however, many of the established computing codes available today will not run well or efficiently, since they were written in the era of single processor systems, or for a very low degree of parallelism. Re-writing these

---

[10]CSEM: Community Earth System Model http://www.cesm.ucar.edu/.

[11]IS-ENES: European Network for Earth System Modeling https://verc.enes.org/.

codes will entail great efforts and necessarily involve a choice of new programming languages and software engineering methods. So the all-up combination of model coupling, new scientific knowledge integration, code rewriting with new programming languages and the use of new software engineering methods, will indeed demand immense talent and expertise.

Even the choice of machine architecture is not a simple matter, for today there is a plethora of road-maps to choose from, beginning with commodity clusters of simple PC-like equipment, all the way up to massively parallel systems with millions of processor cores tied to specialized high-bandwidth low-latency interconnection fabrics. It is also possible to mix accelerator cores with regular processor cores within a single system, e.g. CPU-GPU,[12] for the purpose of optimizing performance, and for lowering system energy requirements. Ultra high-end systems are normally *benchmarked* against each other to determine their performance rankings and desirability—as is done in the TOP500 Supercomputing Sites, Green500, Graph 500, and the HPC Challenge Benchmark.

Looking forward to the years ahead however, we can expect that Moore's Law will deliver top performing machines of exaflop level capability, all having many petabytes of solid state memory.

Unfortunately, these machines will also require megawatts of power to run and therefore be limited as to which organizations can afford them. Herein lays the benefit of *Cloud Computing*, a new form of time-sharing that allows expensive machines and storage costs to be charged out among many users.

On the other hand, the work load of an Earth System Simulator with its job of assimilating all available Earth Observation data and staying abreast of the dynamic changes of the many systems of our planet is likely to overwhelm even the most powerful of high-end machines available by the end of this decade, so a network of machines will most likely be needed, or more accurately, a hierarchy of machine types.

Over the horizon however, changes are on the way as follows: normal *bit-reproducible* computing will be enhanced by *probabilistic* designs wherein statistical calculations are enabled from random electronic noise within the processors themselves, instead of the software level. Quantum computing will become feasible in which millions of *state* conditions can be computed simultaneously. System designs will take on more brain-like *neuromorphic* architecture and thereby develop higher integration capabilities. Indeed, adopting these technologies and others, sooner rather than later, we will well be on the way towards a stochastic, multi-exaflop, interactive, immersive 4D Earth System Simulator—which is the key ingredient to *Modeling the Whole Earth System.*

---

[12]CPU-GPU: Central Processing Unit-Graphic Processing Unit.

Microsatellite Ionospheric Network in Orbit

Stuart Eves

Abstract As a consequence of the enormous human and financial costs associated with major earthquakes, a means of providing short-term earthquake forecasting or warning has long been sought. To date, seismic investigations have failed to yield a reliable method of predicting the time, location, and magnitude of impeding events. However, a growing body of evidence suggests that there may be other precursor signatures, including particularly atmospheric effects, which can be utilised to provide some degree of foreknowledge.

This paper reviews the current evidence for the existence of such precursor effects; attempts to evaluate their suitability for providing appropriate warnings; and then proposes a microsatellite constellation concept which could be used as part of an operational system to detect them.

1 Earthquake Risk

Of all natural hazards, earthquakes are perhaps the most devastating from a human perspective. Over the course of the 20th century it has been estimated that earthquakes caused on average more than 20,000 deaths per year. Since one third of the world's population lives in regions considered to be at risk, it is perhaps unsurprising that the damage to infrastructure, even from single earthquakes, has been estimated at hundreds of billions of dollars.

On their own, such figures would appear to justify investment in a forecasting system, but they may well be significant underestimates of the actual risks on a longer timescale. Regions that are subject to frequent earthquake events generally have building regulations that seek to minimise the impact on infrastructure when the tremors strike. These regions also tend to have emergency procedures that help to mitigate the post-event trauma.

But these are generally regions on natural boundaries between tectonic plates where the earthquake risk is well-understood. Potentially more dangerous are the

S. Eves (✉)
Lead Mission Concepts Engineer, Surrey Satellite Technology Limited, Tycho House,
20 Stephenson Road, Surrey Research Park, Guildford, Surrey, GU2 7YE, UK

F. Freund, S. Langhoff (eds.), *Universe of Scales: From Nanotechnology to Cosmology*,
Springer Proceedings in Physics 150, DOI 10.1007/978-3-319-02207-9_13,
© Springer International Publishing Switzerland 2014

inter-crustal earthquakes that affect locations well away from plate boundaries on a far less frequent basis. Examples of this class of event are the magnitude 7+ earthquakes that have rocked mainland China over centuries and the series of magnitude 8 earthquakes that affected the New Madrid region in the US in 1811/1812. The New Madrid earthquakes were powerful enough to change the course of the Mississippi river, a fact which demonstrates the significance of the effects which can be generated.

2 Earthquake Precursors

A number of precursor mechanisms have been suggested that may be diagnostic of a forthcoming earthquake event. These include:

- Emission of radiation in the ultralow frequency region of the electromagnetic spectrum
- Emission of broadband low frequency radio waves
- Emission of radiation in the thermal infra-red region of the electromagnetic spectrum
- Ionization of the air at the ground-to-air interface
- Bursts of light out of the ground known as earthquake lights
- Emission of radon gas from the surface of the Earth
- Changes to the total electron content of the ionosphere

Though there is, as yet, no general consensus in the scientific community on the physical processes that are involved in generating these diverse pre-earthquake phenomena, help has come over the course of recent years from a somewhat unexpected direction: from solid state physics.

While Minoru Freund was working in the mid-1980s at the ETH Zürich on his Ph.D. thesis dedicated to organic superconductors, Alex K. Müller and G.J. Bednorz of IBM Rüschlikon near Zürich published their paper of the high T_c copper oxide-based superconductors, for which they would receive within a year the Nobel Prize in physics. Alex Müller was Adjunct Professor at the ETH Physics department and Mino worked with him, trying to gain a deeper understanding of the electronic states associated with the oxygen atoms that seemed to be crucially important for the electrical conductivity behaviour of superconducting ceramic oxides. However, subsequently Mino's work took off in another direction, namely to address the fundamental question why, upon heating, the electrical conductivity of magnesium oxide, a model insulator, would increase by many orders of magnitude in a distinctly stepwise fashion. This led to a fruitful collaboration between Mino and his father Friedemann Freund and to insight about the nature of electronic defects that are associated with the oxygen anions in oxide materials.

Soon this work led to the recognition that the same type of electronic defects are also present in silicate minerals and by the mid- to late 1990s it become clear that the same electronic charge carriers can be activated by stress applied to rocks. Since the rocks in question make up the bulk of the Earth's crust in the seismogenic zone at the depth of about 5–35 km depth, this insight marked the beginning of a new era of studying stress-activated electrical processes in rocks and their role in producing pre-earthquake signals.

In order to provide a reliable earthquake forecasting service it is clearly necessary to identify the precursors which are associated with most, possibly all, earthquake events. If they occur sufficiently in advance of the actual event, they provide a useful warning period. A good warning service would also provide an estimate as accurate as possible of both the time of the earthquake and of its magnitude.

2.1 Infra-Red Radiation

There is increasing evidence from satellite measurements of infra-red radiation in association with earthquakes. One example is shown in Fig. 1, which shows data based on satellite measurements made by the AVHRR instrument on one of NOAA's polar orbiting meteorological satellites.

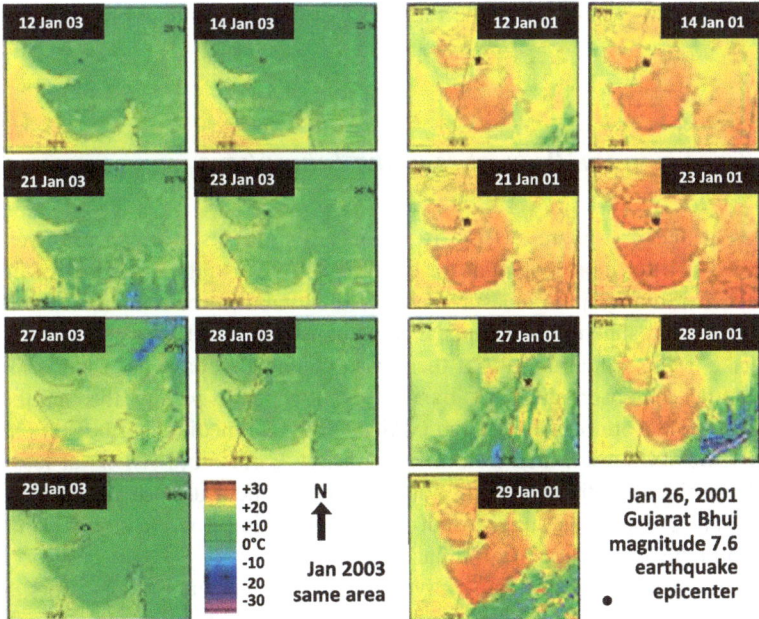

Fig. 1 Satellite detection of presumed pre-earthquake thermal infrared anomalies in India (after Saraf et al. [5]): Land Surface Temperature (LST) maps of Gujarat region for the time period 21 Jan. to 29 Jan. during a "normal" year (2003) and during 2001, when the magnitude 7.6 earthquake occurred. Maximum of the thermal infrared (TIR) anomaly was Jan. 23, 2001

The effects were at a measurable level 12 days before the earthquake event itself, suggesting that such satellite-based IR measurements may be of some utility for earthquake forecasting. Apparent changes of between 2 and 10 K have been claimed.

There are differing hypotheses concerning the origin of the IR signature illustrated here. Freund et al. [3] have experimentally demonstrated a spectroscopically distinct narrow band emission related to the de-excitation of excited species close to the Earth's surface. Other authors prefer a more classical interpretation based on the emanation of gases from the ground and/or moisture condensation in the atmosphere leading to the release of latent heat. Tronin [6] suggests that thermal effects may be measurable for earthquakes of magnitude 5 and greater. However, there is, as yet, insufficient data to confirm whether these effects are real and will occur in association with all rock and surface soil types.

At present it is also uncertain whether such measurements can be made reliably in regions with significant meteorological activity which could impose a "noise" signature onto the data. It is also apparent from Fig. 1 and other results published in the literature that the IR signature is spatially extended and that the peak of the IR intensity does not necessarily coincide with the epicenter of the subsequent earthquake. As a result, it is unlikely that, taken in isolation, IR intensity measurements, i.e. radiative temperature measurements would be able to provide a good indication of the location of a forthcoming earthquake. If and when spectroscopically resolved IR measurements from satellite altitudes become available, IR signatures characteristic of the stress build-up in the Earth's crust might provide a better diagnostic tool.

2.2 Changes in the Properties of the Ionosphere

Changes in the ionosphere, arising from ionisation effects at lower altitudes, are considered to offer more potential for earthquake forecasting. A number of papers have reported tomographic measurements of the total electron content in the lower layers of the ionosphere, which appear to demonstrate correlation with subsequent earthquake events. The majority of the measurements that have been made to date involve either the observation of L-band navigation signals passing through the upper regions of the atmosphere from a GPS satellite in a medium earth orbit (MEO) at around 20,000 km altitude to a mission in low Earth orbit, (LEO) at less than 1,000 km altitude. Estimates of the total electron content derived from variations in the signal propagation suggest that signatures lasting a week or more may be produced in the lead-up to an earthquake event.

These measurements are potentially complicated by a number of factors. Solar storms can also generate variations in the total electron content lasting a number of days, although such variations are likely to be distributed over a larger region than any effects associated with an impending earthquake.

Figure 2 shows the time series of the GPS-derived total electron content variability observed from Feb 23 to March 16, 2011 close to the epicentre of the Tohoku

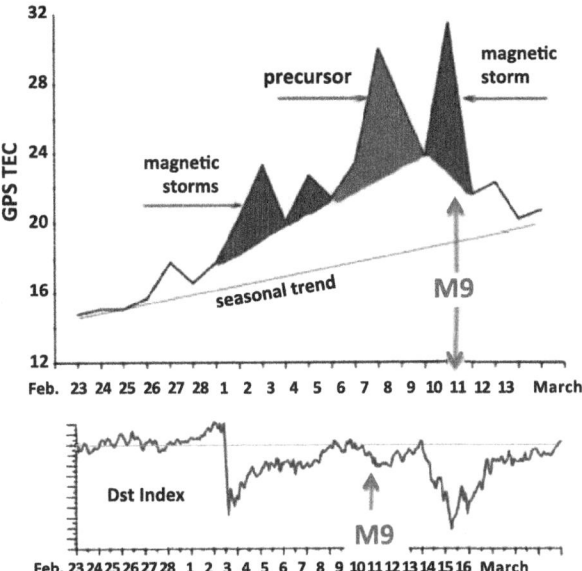

Fig. 2 Total electron content variability and geomagnetic disturbance storm time observed close to the epicentre of the Tohoku magnitude-9 earthquake

magnitude-9 earthquake. The lower panel shows the geomagnetic disturbance storm time (Dst) data provided by the geomagnetism World Data Centre (WDC) in Kyoto, Japan for the same period.

It could, of course, be sheer happenstance that an earthquake event coincides with the peak of the solar-induced geomagnetic disturbance, but there is a possibility that stress-activated charges within the rocks in the Earth crust, which have been hypothesised by Freund et al. [2], would couple to the geomagnetic field disturbances, creating a Lorentz force that might line up with tectonic stress vectors and trigger an earthquake event. The issue here is that geomagnetic storm-induced disturbances are of a comparable magnitude and duration as the precursor events, complicating the recognition task.

The sun drives a diurnal cycle in the highest levels of the Earth atmosphere, which affect measurements made by ionospheric monitoring systems. Some layers in the ionosphere persist through 24 hours, whereas others are present only during the daytime, as illustrated in Fig. 3, and there is also a seasonal periodicity associated with these measurements.

A further complicating factor is that much shorter duration effects may be created in the atmosphere by electromagnetic discharges associated with thunderstorms, such as sprites, elves and jets. In this instance, it may be somewhat easier to differentiate effects caused by sprites, due to their characteristically short durations and distinct polarization, especially if multiple observations can be made. It should be remembered, however, that sprites are particularly associated with the intense thunderstorms created by the thermal convection processes in the atmosphere, which

Fig. 3 The time-variant
layers in the ionosphere

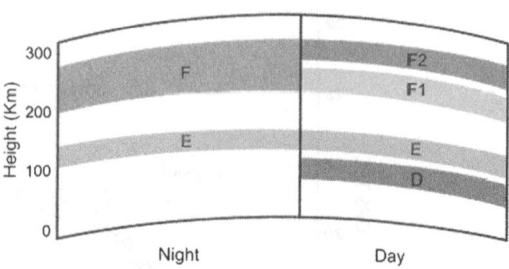

also lead to daily convection rainfall events in the late afternoon local time. Any
monitoring system must take into account these additional sources of diurnal variability.

Occasional scintillation in the atmosphere is a factor that may limit the effective-
ness of this measurement technique over certain regions of the globe. Some of these
scintillations occur at high latitudes, near the poles, and are thought to be driven by
variations in the radiation belts caused by solar activity. Since relatively few people live in the polar regions, this is not a major source of concern. Potentially more
inconvenient is the ionospheric scintillation, which is sometimes seen over low latitude regions of Africa close to sunset. It has been noted that the peak of the sprite
activity occurs in the late afternoon over similar geographic locations, and it may be
that there is a correlation between these two effects.

It is concluded, despite these complicating factors, that sufficient evidence now
exists for variations in the total electron content in the ionosphere associated with
earthquakes to incorporate measurements of this particular parameter as part of the
monitoring system.

2.3 Earthquake Lights

Luminous emissions known as earthquake lights have been documented photo-
graphically in association with some earthquakes. A more detailed discussion of this
phenomenon is available elsewhere in this volume (Chapter by Jorge Heraud), but
the mechanism which creates these effects is only partially understood at present.
What does appear to be true, however, is that luminous effects are observed shortly
before earthquake events, and that the lights are seen largely in association with
igneous rocks. It is concluded that these two facts make earthquake lights an unsuit-
able precursor mechanism for a reliable warning network.

2.4 Low Frequency Radio Waves

The limitations associated with earthquake lights would also appear to apply to the
low frequency RF emissions that have been detected in association with some earthquakes. These signals have been collected by both ground-based sensors and the

Fig. 4 Un-cooled bolometric sensor

French satellite Demeter [1, 4], but only in the hours immediately before the earth-quake event. As a mechanism, which can provide a "red alert" capability, this observable may have some utility, but it is not considered to have significant promise as an "amber alert" warning system.

3 Sensors

Based on the above analysis, the most promising techniques for providing earth-quake warning appear to be monitoring systems which can detect the IR emissions and the total electron content variations in the ionosphere. This section describes the possible sensors required on a space based monitoring network.

3.1 IR Sensors

Historically, IR sensors operating in the thermal infra-red have often required complex cooling systems, but recent developments in un-cooled bolometric sensors have the potential to create an effective monitoring instrument.

An individual sensor could be as illustrated in Fig. 4.

The potential performance parameters for the bolometric detector on which this sensor concept is based are listed in Table 1.

When incorporated in a sensor with a unit length of approximately 14 cm, and a mass of 2 kg, this detector would deliver the following performance:

- Noise equivalent temperature difference (NETD) for a 300 K ground scene = 0.4 K
- Ground Sample Distance = 300 m
- Coverage Swath = 100 km

The effective swath of such an individual instrument from low Earth orbit is clearly too narrow to provide an effective monitoring system. Hence, as with existing disaster monitoring satellites in the DMC network, a multiple aperture sensor system is

Table 1 Bolometric infrared detectors

Detector Array Parameter	Value
Model Name	UL 01 01 1
Manufacturer	ULIS (Grenoble, France)
Detector Type	Microbolometer Detector Array
Detector Material	Resistive Amorphous Silicon
TCR of detector material	2.5 % K^{-1}
Design Waveband	8–14 μm
Pixel Count	240 × 320
Pixel Pitch	45 μm × 45 μm
Fill Factor	>80 %
Sensitive area	11.4 mm × 10.8 mm
Responsivity (mV/K)	4 mV K^{-1}
Peak Responsivity (W/K)	7×10^6 V W^{-1}
NETD @ 300 K w/ f/1 optics	<120 mK
Thermal time constant	4 ms (−3 dB cut-off)
Frame Rate	50–60 Hz (5.5 MHz clock)
Rms noise	480 μV
Dynamic Range	60 K (−10 °C to +50 °C)
Power Consumption	<200 mW
Weight	<50 g
Cost	€10,000

envisaged. In this instance, six similar apertures mounted so as to provide contiguous swaths on the surface of the Earth would provide an effective coverage swath approaching 600 km.

3.2 RF Sensors

Monitoring of the changes to the ionosphere may be achieved by the use of GNSS receivers on a satellite in low Earth orbit (also see Chapter by Kunitsyn et al.). From the perspective of a satellite in LEO, at an altitude of a few hundred kilometres, the satellites in the global navigation constellations (GPS, GLONASS, Galileo, and Beidou) appear to rise and set. By monitoring the propagation of these navigation signals through the upper layers of the atmosphere, tomographic measurements of the total electron content can be obtained. Single-frequency demonstrations of this technique have been performed previously using an array of four GPS antennas mounted on the rear of the Topsat satellite mission (Fig. 5).

In the case of the TopSat mission, only one GPS frequency was monitored. Whilst this allows measurements of received signal strength to be made, an accurate eval-

Fig. 5 Topsat GPS receiver
and antennas

Fig. 6 Small satellite
magnetometer

uation of the total electron content versus altitude is not possible. The receiver envisaged for nanosatellites, called MinoSats in appreciation to the contributions by
Minoru Freund to this emerging field, will simultaneously monitor two GPS frequencies (L1 and L2C), allowing the ambiguities to be resolved. A dual-band receiver system will be flown on the TechDemoSat-1 satellite in 2014, and the mass
of this unit is approximately 1 kg.

3.3 Magnetic Sensors

It has been suggested that measurements of magnetic field variability may also assist
in the determination of variations in the ionosphere. Suitable devices are routinely
incorporated in small satellites as part of the baseline platform design (since they
are required to support the operation of the magnetorquer devices, which form part
of the AOCS design).

The instrument shown in Fig. 6 has the dimensions $99 \times 35 \times 52$ mm and a
measurement sensitivity of ± 10 nT. It weighs just 140 g.

Correlation of the measurements from this device with the other sensors may
provide additional confirmation of changes associated with impending earthquake
activity.

Fig. 7 Satellite design concept, shown in deployed flight configuration and in "stacked" configuration for launch

4 Satellite Design

The envisaged satellite design that would incorporate the sensors described above is illustrated in Fig. 7.

The total mass of this satellite is approximately 50 kg, with a potential payload mass of up to 20 kg. An initial assessment of the total six-aperture IR imager payload is 12 kg. The tomographic RF monitoring system would ideally be equipped with at least four antennas, and is expected to contribute a further one kilogramme to the total payload mass. Redundant magnetometer devices of the type described above would be incorporated into the design as part of the 30 kg platform mass.

In order to provide the necessary coverage, it is important that the satellites have a duty cycle that allows them to collect data whenever they are over the land surface of the globe. The power subsystem is the most obvious limitation on the duty cycle, and the platform illustrated here can provide an orbit average power in excess of 20 W. The power required by each of the IR sensors is 2 W (for a total of 12 W). The GPS receiver needs a total of 4 W. A preliminary power budget would thus appear to indicate an adequate margin to allow the necessary periods of operation.

Further potential limitations on the duty cycle are the memory available on the satellite and the downlink data rate. Increasingly, terrestrial technology developments are making these requirements easier to accommodate. The baseline memory capacity of the platform shown here is 128 Gbyte, and the data downlink rate can be in excess of 100 MB/s if required.

5 Monitoring System

Clearly in order to derive warning information from both the IR and RF sensors, a change-detection processing approach is required. Measurements made at one epoch need to be compared with recent, similar measurements made during ambient conditions. This requirement becomes a driver for the system design, since it is clearly

Fig. 8 Contiguous coverage swaths from 5 operational satellites

essential to have rapid repeat measurements, not only to provide confidence that any possible earthquake precursors will be observed, (ideally more than once), but also to ensure that the false alarm rate is not increased by virtue of having inappropriate previous measurements with which to compare the current observations. Seasonal variations in both the thermal environment and the ionospheric conditions are to be expected, and these must be accounted for by the change detection process.

It will be noted that measurements made by the RF tomographic technique sample the atmosphere at locations which are physically displaced from the nadir-pointing measurements made by the IR sensors. Since the precursor signatures that the system is seeking to measure are thought to originate up to a week in advance of the earthquake event, and to persist up to the time of the earthquake itself, it should be possible to make reasonably reliable predictions based on temporally separated measurements. Ideally though, there should be as close a temporal correlation as possible between the IR and RF measurements.

One possible means of achieving this temporal correlation over regions equipped with suitable receivers could be to incorporate an active beacon payload on the satellite, operating at VHF and UHF frequencies. This approach involves measuring the properties of the ionosphere by comparing its effects on the propagation of signals from the satellite to the Earth's surface. The feasibility of including an active payload of this sort on the satellite described here would critically depend on the power consumption and, hence, the length of time that the transmitters were required to operate.

The solution proposed here is a constellation of 5 operational satellites and an on-orbit spare in a common orbital plane, each of which would notionally provide coverage of a 600 km wide swath on the surface of the Earth as the planet rotates under the orbit plane. The coverage swath pattern that might be expected from a constellation in a high inclination orbit (at an altitude of approximately 700 km, and at an inclination of 60 degrees, or greater) is illustrated in Fig. 8.

A constellation of this size will be capable of providing at least two imaging coverage opportunities per day: one occurring on the ascending passes and a second on the descending passes.

The total mass of the satellites that would be required to populate this single-plane constellation is 300 kg. The nature of the design illustrated in Fig. 7 would allow these satellites to be accommodated on a single dedicated launch vehicle.

It is axiomatic that in order to provide adequate warning of a potential earth-quake, the data would need to be delivered to the ground for processing in a timely fashion. A network of four ground stations, spaced at roughly equal separations in longitude would suffice to achieve a mean data delivery time of approximately three hours.

An alternative orbital configuration would be to utilise a lower inclination orbit at approximately 50 degrees inclination to the equator. The satellites would then not provide coverage of the polar regions, and would be constrained to operate over the regions of the globe where people are actually at risk from earthquake events. As a result of the increased proportion of time spent over populated areas, the timeliness of the system would be improved relative to the polar configuration.

6 Conclusions

The understanding of earthquake precursor signals has advanced to the point where it is reasonable to consider that satellites equipped with IR and RF monitoring sensors could offer the capability to provide "amber alert" warnings up to a week in advance of the event. A constellation of five operational satellites would provide coverage of the populated regions of the globe which are at risk, including those area potentially subject to intercrustal earthquakes.

References

1. J. Blecki, M. Parrot, R. Wronowski, Studies of the electromagnetic field variations in ELF frequency range registered by DEMETER over the Sichuan region prior to the 12 May 2008 earthquake. Int. J. Remote Sens. **31**(13), 3615–3629 (2010)
2. F.T. Freund, M. Lazarus, G. Duma, Top-down and Bottom-up coupling between ionosphere and solid earth. Paper presented at AGU Fall Meeting, Session NH13, San Francisco, CA (2010)
3. F.T. Freund, A. Takeuchi, B.W.S. Lau, A. Al-Manaseer, C.C. Fu, N.A. Bryant, D. Ouzounov, Stimulated thermal IR emission from rocks: assessing a stress indicator. eEarth **2**, 1–10 (2007)
4. F. Nemec, O. Santolik, M. Parrot, J.J. Berthelier, Spacecraft observations of electromagnetic perturbations connected with seismic activity. Geophys. Res. Lett. **35**, L05109 (2008). doi:10.1029/2007GL032517
5. A.K. Saraf, V. Rawat, P. Banerjee, S. Choudhury, S.K. Panda, S. Dasgupta, J.D. Das, Satellite detection of earthquake thermal infrared precursors in Iran. Nat. Hazards **47**, 119–135 (2008). doi:10.1007/s11069-007-9201-7
6. A.A. Tronin, Remote sensing and earthquakes: a review. Phys. Chem. Earth **31**(4–9), 138–142 (2006)

Formation Flying, Cosmology and General Relativity: A Tribute to Far-Reaching Dreams of Mino Freund

Douglas Currie, James Williams, Simone Dell'Agnello, Giovanni Delle Monache, Bradford Behr, and Kris Zacny

Abbreviations

Apollo	refers to the NASA missions to the moon of 1969–1974
APOLLO	Apache Point Observatory Lunar Laser-ranging Operation
CCR	Cube Corner Reflector
CMB	Cosmic Microwave Background, Core-Mantle Boundary
CoM	Center of Mass
ESA	European Space Agency
FF	Formation Flying
GLXP	Google Lunar X Prize
GRACE	Gravity Recovery and Climate Experiment
GRAIL	Gravity Recovery and Interior Laboratory
IAS	Italian Space Agency
INFN	Istituto Nazionale di Fisica Nucleare
INFN-LNF	INFN-Laboratori Nazionali di Frascati
ISAR	Interferometric Synthetic Aperture Radar
JPL	Jet Propulsion Laboratory
LLR	Lunar Laser Ranging
LLRRA-21	Lunar Laser Ranging Retroreflector Array for the 21st Century
LUNAR	Lunar University Network for Astrophysical Research

D. Currie (✉) · B. Behr
Department of Physics, University of Maryland, College Park, College Park, MD, USA

J. Williams
Jet Propulsion Laboratory, California Institute of Technology, Pasadena, CA 91109, USA

S. Dell'Agnello · G. Delle Monache
Istituto Nazionale di Fisica Nucleare, Laboratori Nazionali di Frascati, Frascati, Italy

K. Zacny
Honeybee Robotics Spacecraft Mechanisms Corporation, Pasadena, CA 91103, USA

F. Freund, S. Langhoff (eds.), *Universe of Scales: From Nanotechnology to Cosmology*,
Springer Proceedings in Physics 150, DOI 10.1007/978-3-319-02207-9_14,
© Springer International Publishing Switzerland 2014

NASA   National Aeronautics and Space Administration
SCF   Space Climatic Facility
UMCP   University of Maryland at College Park

Mino had a wondrously wide range of interests and projects. We would like to address three areas that will carry into the future some of Mino's dreams, his concept of swarms of satellites flying in formation, observing the dark un-observed domain of the past universe and the testing of General Relativity involved in the fundamental inconsistency of General Relativity and Quantum Mechanics—the ultimate in the connection of the macro to the micro scales of the physical universe.

1 Formation Flying (FF) and the RHP R-Sensor

Mino addressed a variety of programs involved with multiple satellites, either flying in a fixed formation and/or in a swarm. We will discuss the R-Sensor, an instrument for maintaining such a formation.

1.1 RHP R-Sensor

Radio-Hydro Physics LLC (RHP) has developed a unique instrument that can be used to measure the range (or distance) and the range/rate between the instrument and other locations that are marked with a Cube Corner Reflector (CCR). In particular, this instrument, as seen in Fig. 1, can be used to determine the range and range rate between two satellites. A laboratory model of the "R-Sensor", as seen in Fig. 2, has been built to demonstrate the capability of this technology. In a field demonstration, the R-Sensor has shown a precision in measuring the range of 20 picometer (r.m.s.). This precision was obtained with one second integration. For the range/rate, in one second, one obtains 40 picometers/sec. The current effort is addressing the development of a design for flight implementation.

1.2 Formation Flying about the Moon

One of the objectives of this technology is to support formation flying, that is, missions in which multiple satellites fly in a fixed spatial configuration. An example of such a mission was NASA's Gravity Recovery and Interior Laboratory (GRAIL) mission [50]. The primary objective of the GRAIL mission was the high accuracy mapping of the gravity field of the moon. In turn, this map provides a view into the crust and upper mantle of the moon. The GRAIL mission operated by measuring the distance between two sub-satellites flying in "train" formation, and using a

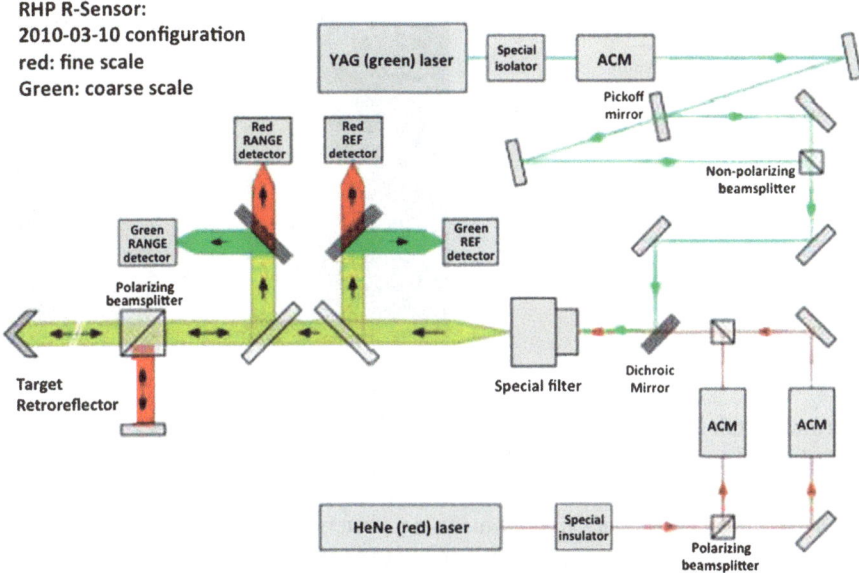

RHP R-Sensor:
2010-03-10 configuration
red: fine scale
Green: coarse scale

Fig. 1 Layout of the RHP R-Sensor, proposed for measurement of inter-satellite range and range rate control of satellite formation or cluster flying

Fig. 2 Operational Implementation of the RHP R-Sensor. This has been used to evaluate the measurement accuracy of the R-Sensor

Fig. 3 Artist's conception of the GRAIL mission, showing the measurement link with a *thin red line* and the communication links with a *thick green line*

Fig. 4 The gravity map
obtained by the GRAIL
mission, revealing the various
density features beneath the
surface

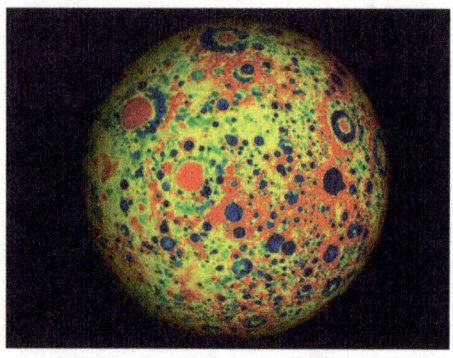

microwave link to measure the distance between the two sub-satellites with an accuracy of about one micron (Fig. 3). By measuring and then analyzing the differential motion of the two satellites, one can determine the local gravity field.

GRAIL detected impact basins, mascons, craters of various sizes, mountains, crater central peaks, and volcanic structures. GRAIL was sensitive to the density, density variations, and thickness of the crust [43] (Fig. 4). It found subsurface structures that appear to be ancient dikes [2]. Spacecraft with low non-gravitational accelerations using optical range and range/rate rather than microwave, as in the R-Sensor, could improve this technique by several orders of magnitude.

By making a detailed comparison of the GRAIL gravity field and the altimetry that has been obtained from Lunar Reconnaissance Orbiter (LRO) [40] it is found that at small scales most of the gravity field variation is due to topography [50]. The remaining field at large and small scales results from sub-surface density and structure variations. For example, the volcanic complexes such as domes are being detected [23].

1.3 Future for Precision FF

The role of the next generation of precision formation flying will now be addressed. The simplest are the next generation GRAIL and GRACE missions. The latter is essentially the same as GRAIL, except that it is investigating the gravity field of the earth rather than the moon. Further downstream, satellite formations may be used for radio and optical interferometry for astrophysics. The control of such formations is more challenging, since one cannot rely on a pair of satellite flying "in train" as in the case of GRAIL. More accurate Interferometric Synthetic Aperture Radar (InSAR) systems would address the ground motion that in turn addresses volcanic inflation and earthquake stress analysis. However, the control of such an expanded constellation, of the type that has been studied for the US Naval Observatory, allows such interferometry to be performed throughout the orbit.

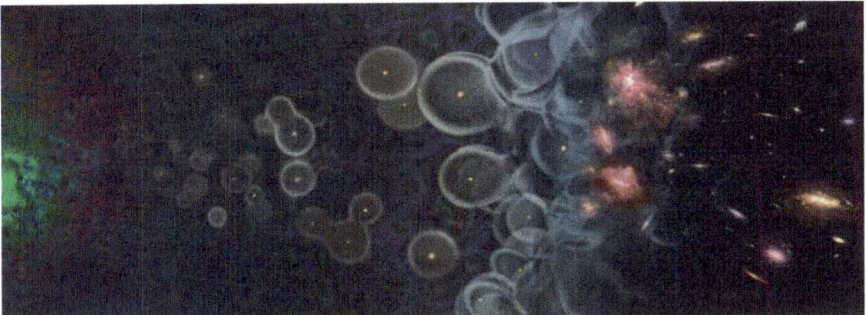

Fig. 5 Schematic of the evolution of the Universe, showing the formation of Cosmic Microwave Background (CMB) on the *left*, the essentially neutral Dark Ages before any stars, the formation of the first stars, the overlap of their ionized regions leading to the re-ionization of the Universe, and the essentially ionized, highly structured modern Universe on the *right*. The large bubbles illustrate the "Stromgren spheres"—large regions that are ionized by the extreme ultraviolet radiation from the young early stars

1.4 Role of R-Sensor for Precision FF

In the case of more complex formations, that is, formations in which satellites are not all flying in a "train" that follows the same orbit, there is a much greater issue. We have investigated such a case and compared it with the "nominal" procedure of using differential GPS to prevent the loss of the operational configuration and possible collisions. In this case, the operation with the R-Sensor reduced the required delta V for configuration maintenance by a factor of ten. This means that the required rate of use of the station-keeping fuel is reduced by a factor of ten and in turn the lifetime of the component of the formation is increased by a factor of ten, resulting in a great reduction in cost of maintaining the formation.

2 Cosmology of the Early Universe

Next we would like to briefly address another of Mino's interests—Cosmology and the investigation of some of the unknown domains in the history of the universe.

The "Lunar University Network for Astrophysics Research" or LUNAR consists of various University and Research Centers that are funded by the NASA Lunar Science Institute, a "Virtual" Institute at located at the Ames Research Center [7]. Our team is led by Jack Burns of the University of Colorado. Our LUNAR team has three foci, the Cosmic Dark Ages, tests of General Relativity and Heliophysics. Although I shall concentrate on our work in testing General Relativity with Lunar Laser Ranging, let us first briefly describe the Dark Ages project.

The science objective of the Cosmology key project within LUNAR is the so-called "Dark Ages", that is, the time after the Cosmic Microwave Background (redshift of 17) until the deepest domain that can be reached by the Hubble and Webb space telescopes (redshift of ~10). Figure 5, using a logarithmic time scale shows

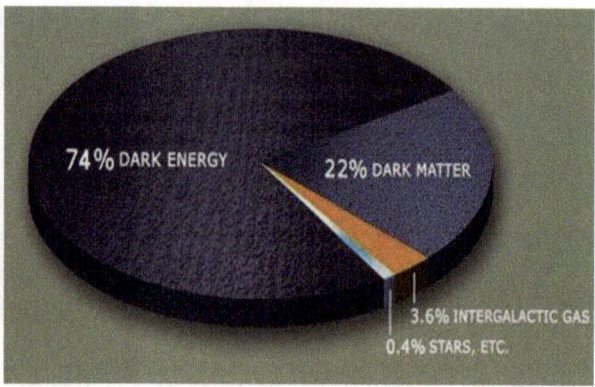

Fig. 6 Illustration of the various major content of the current Universe. This illustrates the importance of the Dark Ages and the understanding the relation between General Relativity vs. Quantum Mechanics

the Big Bang on the left, followed by the Dark Ages where the first stars are formed, the first proto-galaxies are formed and the first black holes are formed [21]. Currently, our knowledge of these events is only via theoretical models [32]. The roles of Dark Matter and Dark Energy in the formation of our current Universe are purely hypothetical. The region is denoted "Dark" both because of our lack of knowledge and because there is no light emitted from much of the region that reaches us for analysis [25]. The investigation of what happened in this region is particularly important with respect to understanding the role of Dark Matter, which with Dark Energy comprises almost all of the content of the Universe as indicated in Fig. 6.

How do we probe this region, where there is either no light that can escape or too faint light for any current telescopes? It is a region filled with atomic hydrogen, and atomic hydrogen emits and absorbs the spin-flip transition at 21 cm or 1.4 GHz. Thus we could determine temperatures and temperature distributions by observing the 21 cm radiation. But this radiation is red shifted down to the tens of megahertz where the radio interference from earth is horrendous. The only place in the solar system that is sufficiently quiet to investigate these dark ages is the far side of the moon.

Both a near term and a long term approach are under investigation. The near term is a Discovery mission—DARE. This consists of a lunar orbiter with a spectrometer (Fig. 7) that observes the all-sky radiation when the lunar orbiting satellite is over the far side of the moon [5, 6]. This proposal, generated by Jack Burns of the University of Colorado and his team at the Ames Research Center, did not make the cut on the recent competition, but appears to be very probable for the next go-around. The long term solution is an array of interferometers on the far side. One approach is to put metal coatings on a Mylar-like material to make the antennas and robotically spread this out on the lunar surface (Fig. 8). This will allow evaluation of the spatial power spectra of the red-shifted 21 cm radiation, which will address the evolving structure of the ionization in this region.

Fig. 7 Artist's conception of the DARE satellite

Fig. 8 Artist's conception of the microwave antennas on the far side of the moon

As discussed, it is extremely difficult to observe this radiation on earth, both due to the ionosphere blocking the lower frequencies and radio interference for earth bound transmitters. Thus large radio antennas on the far side of the moon look extremely interesting [7]. There is a negligible ionosphere on the moon, and on the far side of the moon the emissions from earth's radio transmitters, TV stations, orbiting satellite emissions, etc. are blocked by the body of the moon. Such an observatory would address such questions as: What is Dark Energy and how does it evolve in time? Were there "exotic" heating mechanisms, such as Dark Matter decay, that occurred before the first stars formed? And how did matter assemble into the first galaxies, stars, and black holes?

3 Lunar Laser Ranging: Relativity and Gravity

Now I wish to address an area that is a little closer to my heart, that is, Lunar Laser Ranging, past, present and future. In particular, we want to discuss our "Next Generation" approach to understanding Gravity and Relativity [44, 46]. As mentioned earlier, this is currently supported as a key project of LUNAR headquartered at the University of Colorado and supported by the NASA Lunar Science Institute at Ames Research Center [31]. The research on the Next Generation Retroreflector was previously supported by a contract at the University of Maryland, College Park in the NASA Lunar Science Sortie Opportunities (LSSO) program.

But first a little background:

Why should we care about new measurements of gravity? For the past five centuries Gravity has been the central feature in our understanding of the external universe in what is now called astrophysics. Tyco Brahe, Kepler, Newton and Einstein—all these great minds have described and analyzed both observations and theory describing gravity. Therefore, we should now know all about it—Right?

Wrong—Today as we look out to the universe, the matter known to Newton and Einstein makes up less than 1 % of the universe. Dark Matter and Dark Energy dominate as illustrated in Fig. 6 and we have yet to understand them. Even worse, in the last century Quantum Mechanics has accurately explained the phenomena of the small and has been tested with phenomenal accuracy. And yet we know that Einstein's General Relativity and Quantum Mechanics cannot both be correct. So we must keep pushing on more tests of General Relativity, testing to the limit of available technology.

3.1 Brief History of Lunar Laser Ranging

Back in days of the Apollo Missions, a group of us, initially centered under Robert Dicke at Princeton University and then led by the University of Maryland, College Park, investigated the possibility of using laser ranging to the moon to address critical questions in General Relativity and Lunar physics. A concept using retroreflectors was developed and analyzed [7, 10]. This special set of mirrors was then carried to the moon and deployed by the astronauts [1, 4]. The "Lunar Laser Ranging Observatory" was developed at the McDonald Observatory in Texas [38] to fire short laser pulses to the mirrors, which then send the light back to the observatory. By timing the interval between transmission and return we were able to determine the distance with an uncertainty of ~300 mm. While all the other Apollo experiments left on the moon required power and were shut down after a few years, these retroreflectors are still operating and ranging continues to this day to generate new discoveries about gravity and lunar physics.

3.2 Operational Procedure

The process of Lunar Laser Ranging (LLR) consists, first, of placing on the lunar surface retroreflectors as the one shown in Fig. 9, that is, retroreflectors that can receive a laser pulse from the earth and send it directly back in the same direction with no delay [27]. Thus it can be very effective in assuring a useful signal level on earth. At an observatory on earth, a laser system transmits a short pulse in time that is spatially coherent. This narrow laser beam is then collimated and pointed with a telescope system. Initially these were large astronomical telescopes, but today they may also be smaller telescope dedicated to satellite and lunar ranging. This short

Fig. 9 The retroreflector
array as placed on the lunar
surface by Neil Armstrong
during the Apollo 11 Mission

pulse (<1 nanosecond, where nanosecond of delayed pulse return represents 15 centimeter of range) is then reflected at the moon. A portion returns to earth and is collected by the telescope. This time interval of ∼2.5 seconds between transmission and return is precisely measured and stored for processing.

These measurements are taken frequently. Originally, three times a day, but a lower rate today. This time series of measurements is then compared to the current best model of the orbit and librations of the moon. The orbit and physical libration information has been developed by the Jet Propulsion Laboratory (JPL) and accounts for the effects of the other planets and other bodies in the solar system [11, 39, 49]. The differences between the measured and modeled ranges (a.k.a. the residuals) are then analyzed to address the various signatures that are seen in the time series. Specific signatures can be connected with the parameters entering the model and used to adjust the orbital and librational parameters to obtain a more accurate model, reduce the magnitude of the residuals and provide more accurate values for the physical parameters [33, 41]. Thus the new values of the parameters related to General Relativity and the structure of the moon are determined.

3.3 What Has the LLR Program to the Apollo Arrays Accomplished?

Some of the results are that gravitational energy has the inertial properties of mass, that the change of gravity over the past four decades project to the past to show that the change since the big bang is less than 1 %, that there is no spatial change of the gravitational constant between here and the moon, and many other tests of General Relativity have resulted from the analysis of the changing distance over the decades.

3.3.1 Lunar Science

Lunar science information can come from perturbations on the orbit and from the influence on the three-dimensional orientation of the moon, the physical librations.

We shall discuss solid body tides (about 10 cm in magnitude), dissipation at and shape of the core-mantle boundary, rotation normal modes, orbit evolution, and a search for lunar interior effects [24, 33].

3.3.2 Tides

There is elastic information from the Apollo seismometers, but that information does not extend to the lower mantle and core. Of the three Love numbers, l_2 is least sensitive to the deep zones so we solve for k_2 and h_2 while fixing l_2 at a model value of 0.0107. Solutions give $k_2 = 0.0241 \pm 0.0020$ [48] and $h_2 = 0.045 \pm 0.008$.

The GRAIL mission provided information on the Love number k_2. Data analysis provides a value with a 1 % uncertainty [48].

3.3.3 Dissipation from Tides and Core

There are many small perturbations on the orientation of the lunar orbit and equator planes, but there is one big effect due to dissipation. Key to separating the two causes of dissipation has been the detection of small physical libration effects of a few milliarcseconds size. Guided by semi-analytical theories for tide and core dissipation [45], we solve for periodic terms in longitude physical librations at 1 yr (annual mean anomaly), 206 d, and 1095 d (1/2 period of argument of perigee) in addition to a tidal time delay and the fluid core K_v/C. The tidal time delay and the core-mantle boundary (CMB) dissipation are both effective at introducing a phase shift in the precessing pole direction. The solution gives dissipation from the core-mantle boundary and tides. Both are strong contributors to the $0.27''$ offset of the precessing rotation pole from the dissipation-free pole, equivalent to a $10''$ shift in the node of the lunar equator on the ecliptic plane [24].

3.3.4 Oblateness of the Fluid Core

Detection of the oblateness of the fluid-core/solid-mantle boundary (CMB) is independent evidence for the existence of a liquid core. In the first approximation, CMB oblateness influences the tilt of the lunar equator to the ecliptic plane. Parameters for CMB flattening, core moment of inertia, and core spin vector, are introduced into torque T_{cmb} in the numerical integration model used for lunar orientation and partial derivatives. Equator tilt is also influenced by moment-of-inertia differences, gravity harmonics and Love number k_2, solution parameters affected by CMB oblateness. Solutions can be made using the core and mantle parameters.

Torque from an oblate CMB shape depends on the product of the fluid core moment of inertia and the CMB flattening, $fC_f = (C_f - A_f)$, where the pole and equator fluid core moments are C_f and A_f. Both are uncertain and there is no information about flattening apart from these LLR solutions. The LLR solution gives $f = (C_f - A_f)/C_f = (2.5 \pm 1.4) \times 10^{-4}$ [47, 48]. For a 370 km core radius the flattening value would correspond to a difference between equatorial and polar radii of about 90 m with a large uncertainty. The f uncertainty seems to imply weak detection at best, but the derived oblateness varies inversely with fluid core moment, as expected theoretically, so a smaller fluid core corresponds to a larger oblateness value. The product $fC_f/C = (C_f - A_f)/C = (1.7 \pm 0.5) \times 10^{-7}$ is better determined than f alone. Core flattening appears to be detected and the foregoing product is more secure in a relative sense than the value of f itself. In the solution the corrections to core moment and CMB flattening are from the DE430 ephemeris [47, 48].

3.3.5 Free Librations

The differential equations for lunar rotation have normal modes, three for the mantle and one for the fluid core. Dissipation has been recognized by LLR from both tidal flexing and the fluid/solid interaction at the core/mantle boundary. Dissipation introduces a phase shift in each periodic component of the forced physical librations. It might be expected that the free physical librations associated with these normal modes would be imperceptible since the damping times are short compared to the age of the Moon.

However, substantial motions are found for two of the modes [8, 9, 22], and we have to ask what is the source of stimulation? Reported here are results from the recent effort with Rambaux that analyzed the DE421 numerically integrated physical librations. The free physical librations depend on the initial conditions for the Euler angles and spin rates, which are adjusted during the LLR fits. The integrated Euler angles were fit with polynomials plus amplitudes and amplitude rates for trigonometric series. More than 130 periodic terms were recognized in two latitude libration angles, while longitude libration yielded 89. The free libration terms were identified among many forced terms.

The longitude mode is a pendulum-like oscillation of the rotation about the (polar) principal axis associated with moment C. The period for this normal mode is 1056 d = 2.89 yr and the amplitude is 1.3″ (11 m at the equator). The damping time is 2×10^4 yr. The lunar wobble mode is analogous to the Earth's polar motion Chandler wobble, but the period is much longer and the path is elliptical. Observed from a frame rotating with the lunar crust and mantle, the rotation axis traces out an elliptical path with a 74.6 yr period. The amplitudes are 3.3″ × 8.2″ (28 m × 69 m). The computed damping time is about 10^6 yr. The two remaining free modes are retrograde precession modes when viewed from a nonrotating frame in space. The mantle free precession of the equator (or pole) has an 81 yr period. An amplitude of 0.03″ is found for this mode, but there is uncertainty because the LLR fit for the integration initial conditions appears to be sensitive to the lunar interior model.

The expected damping time is 2×10^5 yr. The fluid core free precession of the fluid spin vector has an expected period >100 yr; it would be 300 yr for the DE430 integration. The period depends on the CMB flattening previously discussed under Core Oblateness. Based on the trigonometric analysis, this mode must have a small amplitude.

3.3.6 Search for a Solid Inner Core

It is reasonable to expect that the Moon would have a solid core interior to the fluid core, but it remains undetected [24]. The phase diagram for Fe-FeS shows that cooling of fluid alloys of iron and sulfur would freeze out part of the iron while concentrating sulfur compounds in the fluid [42] have a possible seismic reflection from a 240 km radius solid inner core. An inner core might also be detected through its influence on physical librations or gravity. Confirmation and information are needed for this last major unit of the Moon's structure [24].

Lunar Laser Ranging is sensitive to small effects in the lunar physical librations due to lunar structure and interior properties. Predicting the size of inner core effects depends on a number of unknown parameters including the inner core moment of inertia and gravity field, and the mantle's gravity field interior to the CMB. An inner core might be rotating independently or it might lock to the mantle rotation through gravitational interaction. The inner core and mantle interact through their non-spherical gravity fields. This gravitational interaction is expected to be very much stronger than torques from the fluid core so we assume that the mean rotation rates of mantle and inner core are the same. The inner core also interacts gravitationally with the Earth. Like the mantle, the orientation of the inner core is expected to precess at the same node rate as the mantle, but the equator of the inner core is not necessarily aligned with the mantle's equator. The tilts between the two equators and the ecliptic plane will be different and this difference will cause a small variation in the external gravity field of the Moon that might be detected by spacecraft. A strong gravitational interaction between inner core and mantle tends to align their equator planes and a very weak interaction makes the orientations more independent. The inner core rotational dynamics has a resonance, if a precession-like normal mode frequency of the inner core matches the forcing frequency of $-1/18.6$ yr. Close to such a resonance the two orientations could be very different. There are other forcing frequencies that can also resonate causing potentially observable effects in the physical librations. The frequency of the precession-like normal mode would determine which physical libration terms would get modified most strongly.

An inner core can also modify the physical librations in longitude. There are a large number of forcing terms for longitude librations. The inner core introduces a new longitude libration normal mode with a natural frequency and that frequency is a resonance that determines which longitude libration periodicities are most strongly affected. The period of the longitude normal mode might be from less than one year to decades.

To look for inner core and other geophysical effects, the postfit LLR residuals from 1970–2010 for each retroreflector array have been analyzed to produce spectra. The Apollo 11 and 14 arrays are near the equator, so they will be most sensitive to longitude librations. The Apollo 15 array, well north of the equator with a small longitude, provides the most sensitivity to latitude librations. The Lunokhod 2 array is sensitive to both longitude and latitude librations, but the small number of observations (3 %) gives this array the noisiest spectra.

All of the spectra are highest for periods longer than a year. The Apollo 11 spectrum is 9 mm high on either side of the 1056 d mantle resonance. The Apollo 14 spectral amplitudes are highest at 8 mm also near the 1056 d resonance. The amplitudes for the Apollo 15 spectrum are all smaller. The Apollo 15 reflector has the most observations.

Any detection of and information on the Moon's inner core will be a major accomplishment for any technique. There may also be other poorly known geophysical effects. There will be future LLR investigations.

3.3.7 Orbit Evolution

Dissipation in the Moon and Earth causes slow changes in the lunar orbit. The semi-major axis and eccentricity increase with time and the inclination decreases. Dissipation in the Moon also deposits heat in the Moon. This is a minor effect now, but could have been much more important when the Moon was closer to the Earth. Here we summarize the orbit changes.

Table 1 presents dissipation-induced secular rates for mean motion n, semi major axis a, eccentricity e, and the Earth rotation rate ω. LLR results on three lines are compared with model computations on four. Except for the anomalous eccentricity rate, the LLR-determined values are computed from ephemeris DE421 parameters. The LLR integration model for terrestrial tidal dissipation uses Love numbers and time delays for three frequency bands: zonal (long period), diurnal, and semidiurnal. For DE421 the three Love numbers and the zonal time delay were set to model values. The diurnal and semidiurnal time delays were fit to LLR data in creating DE421. For the Moon, the lunar Love number k_2, tidal time delay, and CMB dissipation parameter K_v/C were fit for DE421. For the LLR fits, the Earth tide parameters are sensed through the orbit changes, but the lunar Love number, time delay and CMB dissipation are mainly determined from the physical librations. The "anomalous" eccentricity rate is not present in the DE421 integration, but an additional eccentricity rate is found in solutions using DE421. For comparison, model values of dn/dt, da/dt, de/dt and $d\omega/dt$ were computed for the Earth tides based on the IERS Conventions [28] for the main body and FES2004 results for the ocean tides [26, 35]. There is some uncertainty in converting the terrestrial Love numbers and time delays to orbit rates, but the same theoretical expressions were used for converting the LLR and Earth model parameters and that should minimize differences. The results are presented in Table 1.

Table 1 Dissipation-induced rates for mean motion, semi major axis, eccentricity, and Earth rotation comparing LLR results to an Earth tide model

	Units	Zonal	Diurnal	Semi-diurnal	Earth sum	Lunar tides	Lunar CMB	Moon sum	Anomalous	Total
LLR dn/dt	$''/\text{cent}^2$	0.12	−3.31	−22.88	−26.07	0.20	0.02	0.22		−25.85
LLR da/dt	mm/yr	−0.18	4.89	33.75	38.46	−0.30	−0.02	−0.32		38.14
LLR de/dt	10^{-11}/yr	−0.03	0.16	1.20	1.33	−0.40	0	−0.40	1.32	2.25
Model dn/dt	$''/\text{cent}^2$	0.12	−3.76	−22.61	−26.25					
Model da/dt	mm/yr	−0.18	5.55	33.36	38.73					
Model de/dt	10^{-11}/yr	−0.03	0.22	1.54	1.73					
Model $d\omega/dt$	$''/\text{cent}^2$	0	−196	−1125	−1321					

In the table note that the total Earth dn/dt from LLR and the Earth model differ by <1 %. An independent LLR analysis for total dn/dt of $-25.858''/\text{century}^2$ [22] gives very good agreement with the DE421 mean longitude acceleration of $-25.85''/\text{century}^2$ given here. The DE421 value corresponds to a 38.14 mm/yr semi major axis rate.

There is less agreement between eccentricity rate from LLR and the Earth model because the LLR solutions mainly accommodate the tidal acceleration dn/dt that very strongly affects the LLR data. Most of the Earth tide de/dt comes from the N2 tide, while for dn/dt the M2 and O1 contributions are larger. For the lunar tides, the component with the anomalistic period is most important for de/dt. Accounting for the 0.4×10^{-11} /yr difference in de/dt from the simple LLR integration model and the more complete Earth model, the remaining eccentricity rate is $(0.9 \pm 0.3) \times 10^{-11}$ /yr, equivalent to an extra -3.5 mm/yr in perigee distance and $+3.5$ mm/yr in apogee distance. The inclination rate is not given in the table since it is computed to be only -1×10^{-6} ''/yr. The predicted Earth spin rate change is given in the last line of the table. In decreasing order, the most important tides for secular rotation acceleration are M2, S2, K1, O1, and N2. The S2 and K1 tides do not cause secular changes in lunar mean motion or eccentricity.

There is no evidence for any anomaly in the tidal acceleration in mean longitude. By contrast, the DE421 anomalous lunar eccentricity rate indicates that the LLR integration model needed improvement. Improvements to the terrestrial tide model have subsequently been made and the anomalous rate was cut in half. Computation of lunar orbit evolution over long times needs a good understanding of the various contributions to the secular rates. Long-time evolution of the orbit is complex because of evolving lunar thermal conditions and changing ocean tides [34].

Fig. 10 Libration pattern, showing the apparent motion of the center of mass of the Earth as viewed from the retroreflector arrays

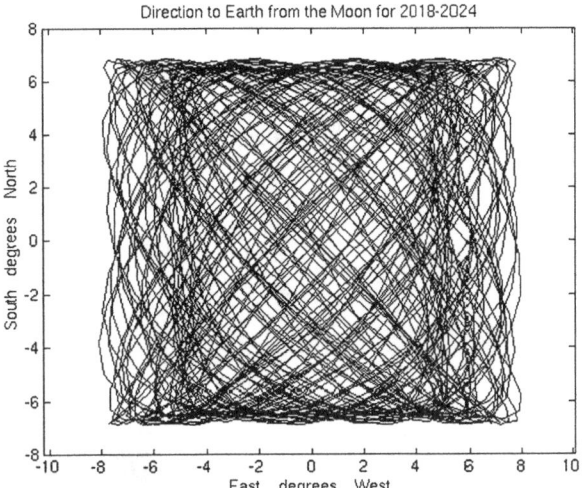

3.4 Optical Libration Problems

As we have seen, the retroreflector arrays left by the astronauts of the Apollo missions have generated a large number of unique new science results. One might ask why the push for a new set of retroreflectors when we are still generating new science with the Apollo arrays. During the past four decades, the laser ranging observatories on the ground have improved measurement accuracy by more than a factor of 200 [29, 30]. As a result, our current limiting accuracy is defined by the Apollo arrays in conjunction with the lunar optical librations rather than, as was the case earlier, the parameters of the ground station. Optical librations are the changes in the apparent direction to the earth as seen from the moon due to orbital eccentricity and inclination. This range of apparent directions to the Earth is illustrated in Fig. 10. The Apollo retroreflectors each consist of panel with 100 or 300 Cube Corner Reflectors (CCRs), each 38 mm. During the monthly optical libration pattern, the angular offset becomes as large as 8° in longitude and 7° in latitude. Thus, as the moon rotates, the panel is tipped with respect to the normal to its direction to the earth. This means that intuitively, we do not know whether a photon was reflected by a CCR at the furthest corner of the panel or the nearest corner of the panel. This results in an r.m.s. uncertainty of 24 mm for the Apollo 11 and 14 retroreflector arrays and 46 mm for the Apollo 15 reflector as depicted in Fig. 11. For unfavorable optical librations the uncertainty can be as large as 70 mm for the last. On a more practical level, the result of the optical libration is to produce a spread in the temporal width of the return pulse, so there is no incentive to install a new laser with a very narrow pulse. Today, the only method of obtained millimeter ranges is to employ a large astronomical telescope to record thousands of returns for a single normal point. This results in fewer observation sessions per month and means that smaller aperture stations cannot achieve the millimeter results. Over the past four decades, the lunar observatories have improved their ranging accuracy by a factor of about 200 as demonstrated by Fig. 11, and yet the fit between observations and

Fig. 11 Residuals between the observations and the best fit model. This indicated the combined uncertainty in the combination of the model and the observations

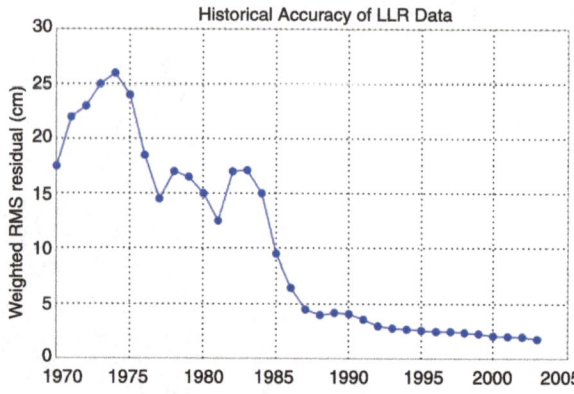

Fig. 12 100 mm Cube Corner Reflector compared to an 38 mm engineering model from the Apollo era

theory has stabilized and not improved over the past decade. To address this NASA has supported our development of the next generation of retroreflector.

3.5 Solution to Optical Libration Problem

The solution to the libration problem consists of the deployment of a single large solid cube corner reflector as shown in Fig. 12 instead of many smaller CCRs whose distance to the earth varies with the change in the librational angles [12–14, 44]. Such an approach should improve our ability to address gravity, relativity and the properties of the moon by a factor of 10 to 100, depending upon the method of deployment on the lunar surface.

3.6 Challenges Involved in Using a Large Solid CCR

While as mentioned, the use of a single large solid Cube Corner Reflector is a theoretical solution to the optical libration problem, there are significant technical challenges to accomplish satisfactory operation. We will discuss the three most important challenges: the fabrication of the CCR, the thermal distortion due to the harsh lunar environment and the stability of the emplacement on the lunar surface.

3.6.1 Fabrication of Large Cube Corner Reflectors to the Required Tolerances

The angular tolerances for the fabrication of the 100 mm CCR are a factor of $\sim$2.5 more difficult than the normal state of the art for the fabrication of the 500 original Apollo CCRs and the thousands of CCRs used in satellites. To illustrate that this is feasible, we have fabricated a 100 mm CCR for which the requirements of an accuracy of 0.2 arc-second was met. On the other hand, we have found that the technology for measuring angles of the back offset faces, critical for proper operation, also has challenges. Extensive data is being collected and discussions with the manufacturer of the interferometers which provide these measurements are proceeding to address this open question.

In addition, for the 100 mm CCR, the homogeneity of the fused silica material is also a challenge. A measurement program is currently underway to address this [18].

3.6.2 Thermal Control to Reduce Thermal Gradients to Acceptable Levels

Thermal gradients within the CCR result in gradients in the index of refraction within the CCR. The later result in degradation of the collimation of the return beam going back to Earth, which in turn reduces the signal level. The harsh environment of the lunar surface, where the temperature can range from 100 K to nearly 400 K means that thermal control is extremely important, far more than in satellites and, due to the larger size, more than for the Apollo arrays.

3.6.3 Emplacement Goal—A Long Term Stability of <100 Microns w.r.t. CoM

The final major challenge addresses the ability to maintain a relatively fixed defined relation between the optical center of the CCR and the center of mass of the moon. This is important since the tests of General Relativity involve the accurate measurement of the motion of the center of mass of the moon, that is, the motion of the moon along a geodesic. The harsh thermal environment of the lunar surface turns this into a challenge. To address it, we consider three different deployment approaches [51]:

Deployment on Lunar Lander The deployment of the LLRRA-21 on a lunar lander is the most likely expectation in the near future. This has the advantage of requiring the minimum of auxiliary equipment and minimizes the required mass for the transport. On the other hand, it suffers from the change in height due to the thermal expansion and contraction of the lander itself. This will limit the accuracy for a single photoelectron return to a few millimeters depending upon the mission. At the same time such an emplacement will allow millimeter ranging by a number of additional stations. This will assure a continuing observation program over the next few decades. In order to reach the millimeter level, one will require ten or more returns to obtain a one millimeter normal point. Such a deployment is being developed with several candidate rides. For example, the Moon Express team is developing a

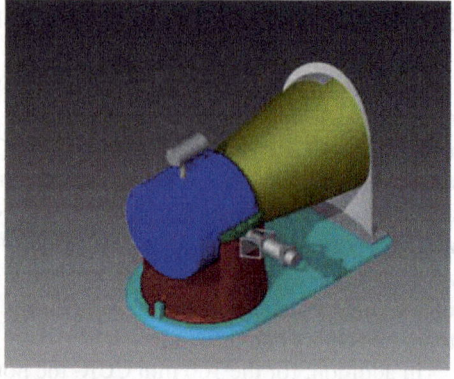

Fig. 13 Artist's conception of the pointing mechanisms and procures to point the package to the center of the earth's librationpattern

Pointing_Titled

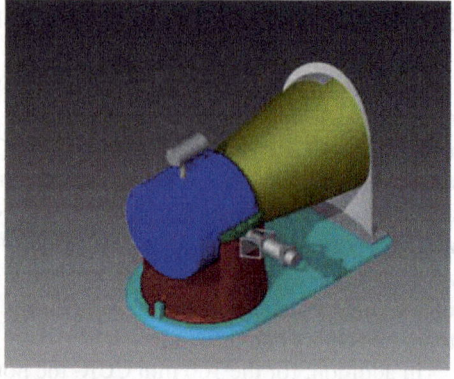

Fig. 14 model of our Lunar Laser Ranging Retroreflector for the 21st Century mounted on the instrument platform of the model of MoonEx1 is shown. In the background are Joe Lazio, Deputy PI of LUNAR, Jack Burns, PI of LUNAR, Doug Currie, PI of LLRRA-21, Bob Richards, COO of Moon Express and Alan Stern, Chief Scientist of Moon Express

lander shown in Fig. 13. Since the LLRRA-21 must be pointed back toward the earth to within 1 or 2 degrees and since the lander landing orientation will not have this accuracy, we need a dedicated pointing mechanism. Such a mechanism is being developed at the University of Maryland. Figure 13 illustrates the conceptual design currently being developed, and Fig. 14 shows a mock-up of the lander with the Development Team.

3.6.4 Deployment on the Lunar Surface

The second method of deployment is directly on the regolith. This will result in a reduced effect of the thermal motion but still be subject to the thermal motion of the regolith, which is a significant portion of a millimeter. This will require a careful thermal design of the support so that it does not contribute to the thermal motion as shown in Fig. 15. This also requires the lander have an articulated arm in order to deploy the surface mounted LLRRA-21. There remains the Day/Night vertical thermal motion of the Regolith of ~400 microns.

Fig. 15 Artist's concept of the current design for the surface deployment possibility. The indicated "tripod" is composed of silicon carbide to achieve a very low coefficient of thermal expansion. In addition, the coupling between the tripod and the LLRRA-21 counteracts the residual coefficient of thermal expansion

Fig. 16 Drilling operation by Jack Schmidt and Gene Cernan which illustrates the challenges of conventional drilling in theregolith

Honeybee Lab Test

3.6.5 Anchor the CCR to the Deep Regolith

In order to escape the problem of the vertical thermal motion of the regolith, we note that at a depth of 0.5 to 1.0 meters, the temperature remains essentially constant during a lunation. Therefore, if we were to anchor the LLRRA-21 to this depth we could deploy the LLRRA-21 in a manner to escape the day to night vertical motion of the regolith. The CCR would then be attached to this deep anchor by a support rod composed of a low thermal expansion material such as INVAR or silicon carbide. However, the process of drilling of the hole for the support rod and anchor would appear to be a non-trivial challenge. During the Apollo mission, drilling was quite difficult (see Fig. 16 with Jack Schmidt and Gene Cernan). This was primarily because previous drilling methods attempted to compress the regolith. In general, the mechanical properties of the regolith strongly resist such compression. However Kris Zacny at Honeybee [20] has developed a technique—pneumatic drilling—in which the support rod is hollow and gas is sent down the hollow core. When the gas exits the hole in the tip, it blows the regolith particles out of the newly formed hole. This technology has already been tested with compacted JSC1a lunar regolith simulant (Fig. 17). Further test have been conducted with the compacted JSC1a in

Fig. 17 Demonstration that
the pneumatic drilling
technique needs only the
weight of the CCR to
excavate a hole in compacted
JSC1alunar regolith simulant

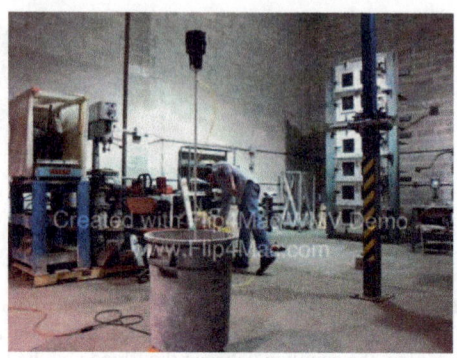

Apollo 17 Spill

Fig. 18 Artist's concept of
the Astrobotics lander
showing the deployment
mechanism on the lander and
the deployed LLRRA-21 in
the anchored mode

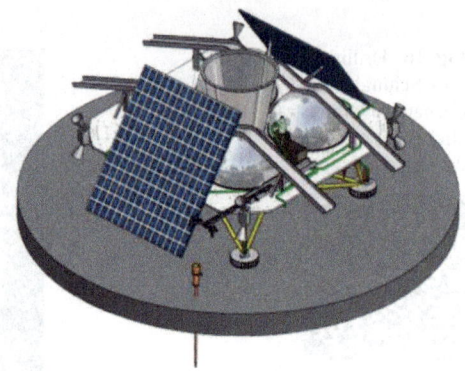

vacuum and at lunar gravity. In order to address the feasibility of implementing such
a drilling technique during an actual lunar landing, Zacny has developed a conceptual design for the pneumatic drill on the lander being developed by the Astrobotics
Team (Fig. 18) [3, 51].

3.7 Current Status of LLRRA-21 Program

3.7.1 LLRRA-21 Design

The preliminary design of the LLRRA-21 package has been completed. This design
is illustrated in Fig. 19. The sunshade, in yellow at the top, blocks the direct sun into
the CCR for most of the lunar day. The CCR is shown in red. Below the CCR are
two thermal shields, to block radiation from the internal surface of the housing from
being absorbed by the CCR and generating thermal gradients. The interior surface
of the inner thermal shield is shaped like the back of the CCR and has a silver
coating to effectively reflect most of whatever solar radiation breaks through the
total internal reflection back to space. Finally, the housing encloses the CCR and

Fig. 19 Artist conception of
the current design of the
LLRRA-21. The details are
discussing in the adjoining
text

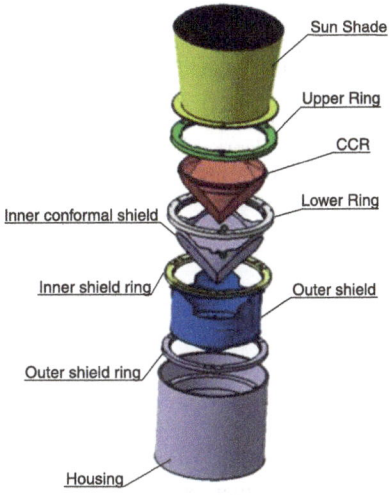

Fig. 20 Prototype of the
LLRRA-21 with stepped
sunshade, housing and 100
mm CCR. This will be used
in the Phase 2 thermal
vacuum testing in the SCF in
Frascati, Italy

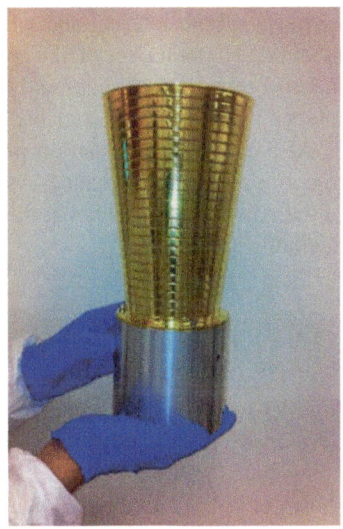

the thermal shields and serves as the interface to the lander/pointing mechanism.
The current prototype is shown in Fig. 20, with the stepped sunshade.

3.7.2 LLRRA-21 Solar/Thermal/Vacuum/Optical Simulation

In order to understand the thermal issues and in order to optimize the performance
a detailed set of programs to simulate the environment has been developed. These
successively determine the deposition of thermal power due to the solar radiation
in the CCR using IDL [RSI] programs developed at the University of Maryland.

Fig. 21 Thermal vacuum
chamber to be used in the
Phase 2 thermal vacuum
testing of the LLRRA-21 at
the new SCF facility
consisting of a large clean
room with two TV chambers

The commercial program Thermal Desktop [C&R Technologies] is then used to convert the power deposition into temperature distributions, while accounting for the solar inputs, the interaction with the regolith as it changes temperature and radiation exchanges, both internal and with space that occur throughout the orbit over a full lunation. The other set of IDL programs convert the temperature distributions into index of refraction changes and finally the optical output and signal level that would be seen by the observatories back on Earth.

At present, the simulation programs are being used to select thermal coatings to optimize the performance. As a portion of this project, the thermal performance of the Apollo arrays is being developed to understand the effects of dust and other degrading processes.

3.7.3 LLRRA-21 Brass Board

In order to understand the details of the challenges of the real hardware, we have developed a full model of the LLRRA-21. This has also provided the basis for developing an assembly procedure. This has been done with flight qualified components, shown in Fig. 21 although it is not the unit that would be expected to fly.

3.7.4 LLRRA-21 Thermal Vacuum Testing

In order to address the validity of the thermal simulation, a series of thermal/vacuum/optical tests have been performed. These have been performed at a special facility (SCF) at the INFN-LNF in Frascati, Italy. This SCF has been created especially for testing retroreflector packages [15, 16]. It has a solar simulator, windows for optical evaluation of the CCR while under test conditions and an infrared transmitting window to allow an infra-red camera to evaluate the temperature distribution while the test is being conducted. The IR measurements allow a cross calibration with thermocouples that are distributed on the package. Figure 22 is a

Fig. 22 The thermal vacuum chamber or SCF at the INFN-LNF where the early prototype of the LLRRA-21 has been tested

Fig. 23 Infrared image of the 100 mm CCR under thermal vacuum test. Temperature variation illustrates the effect of tab conduction and radiation from thin regions

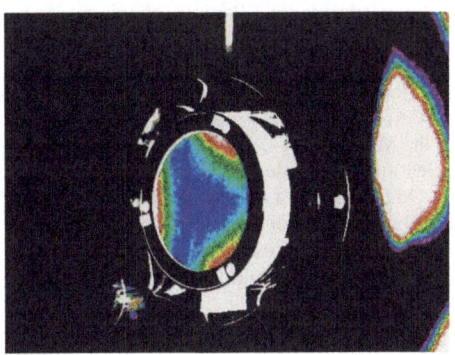

photograph of the SCF in which the previous tests have been conducted. Over the past year, a clean room suitable for testing flight hardware has been developed at INFN-LNF as seen in Fig. 22. Figure 23 is an infrared image of the CCR during one of the test indicating the temperature distribution across the front face of the CCR.

4 But How Do We Get to the Moon?

In response to the Google Lunar X Prize (GLXP) [19] a large number of groups have started plans for a soft lunar landing. Included are several groups that plan a commercial transportation objective. In particular, the current candidates that we are working with are Moon Express and Astrobotics [3]. We have been discussing detailed interface issues with the Moon Express Team for mounting on the lander and, as mentioned earlier, created a design concept for an anchored emplacement. We have a signed agreement with the Japanese group that is considering a retroreflector for SELENE-2 [36, 37]. With Moon Express, located at the NASA Ames Reach Center in Mountain View CA we are planning a landing before the end of 2015.

Working with these commercial groups, Moon Express, Astrobotics and SpaceX is a refreshing experience. They display the youth, enthusiasm and excitement I enjoyed in NASA in the early day.

5 Conclusions

Finally, we wish to present this paper in tribute to the many interests of Mino. This has addressed the control of formations and swarms of satellites and an observational exploration of the unknown aspects of the universe. However, the main focus has been on our program to address measurements addressing gravitation and relativity, fundamental to our understanding the physics of the universe. The latter may contribute to understanding the properties of Dark Matter, Dark Energy and possibly addressing the fundamental conflict between General Relativity and Quantum Mechanics.

Acknowledgements Portions of this research have been supported by NASA Headquarters via the Lunar Science Sortie Opportunities (LSSO) program, by the Planetary Science Division through the NASA Lunar Science Institute to the University of Colorado under the University of Maryland Contract, as well as support by the University of Maryland, College Park. The work of S.D.'A. and G.D.M. is supported by INFN (Istituto Nazionale di Fisica Nucleare, Italy), as part of the MoonLIGHT-2 experiment in the framework of the research activities of the Commissione Scientifica Nazionale n. 2 (CSN2).

Other portions of this research have been supported by the Istituto Nazionale di Fisica Nucleare, Laboratori Nazionali di Frascati (INFN-LNF), Frascati, Italy and the Italian Space Agency (IAS).

A portion of the research described in this paper was carried out at the Jet Propulsion Laboratory of the California Institute of Technology, under a contract with the National Aeronautics and Space Administration.

The first author also wishes to thank Jack Schmidt for discussions and the video.

References

1. C.O. Alley, R.F. Chang, D.G. Currie et al., Apollo 11 laser ranging retro-reflector: initial measurements from the McDonald Observatory. Science **167**(3917), 368 (1970)
2. J.C. Andrews-Hanna, F. Nimmo, J.W. Head, M.A. Wieczorek, W.S. Kiefer, G.J. Taylor, S.W. Asmar, A.S. Konopliv, F.G. Lemoine, E. Mazarico, P.J. McGovern, H.J. Melosh, G.A. Neumann, R.J. Phillips, D.E. Smith, S.C. Solomon, J.G. Williams, M.T. Zuber, Ancient igneous intrusions and the early expansion of the Moon revealed by GRAIL gravity gradiometry. Science **339**, 675–678 (2013)
3. Astrobotics, http://www.astrobotic.com/
4. P.L. Bender, D.G. Currie et al., The lunar laser ranging experiment. Science **182**(4109), 229–238 (1973)
5. J.O. Burns, J. Lazio, Year 3 LUNAR Annual Report to the NASA Lunar Science Institute (2012), eprint arXiv:1204.3574
6. J.O. Burns, T.J.W. Lazio et al., Probing the first stars and black holes with the Dark Ages Radio Explorer (DARE). Adv. Space Res. **49**, 433–450 (2012)
7. Committee on the Scientific Context for Exploration of the Moon, *The Scientific Context for Exploration of the Moon* (National Academies Press, Washington, 2007), pp. 1–120
8. O. Calame, Free librations of the moon determined by an analysis of laser range measurements. The Moon **15**(June-July), 343–352 (1976). Research supported by the Centre National de la Recherche Scientifique of France
9. O. Calame, Free librations of the moon from lunar laser ranging, in *Scientific Applications of Lunar Laser Ranging, Proceedings of a Symposium, Austin, Texas, USA, June 8–10, 1976*, ed. by J.D. Mulholland. Astrophysics and Space Science Library, vol. 62 (Reidel, Dordrecht, 1976), p. 53

10. R.F. Chang, D.G. Currie, C.O. Alley, M.E. Pittman, Far-field diffraction pattern for corner reflectors with complex reflection coefficients. J. Opt. Soc. Am. **61**(4), 431 (1971)
11. J. Chapront, Improvements of planetary theories over 6000 years. Celest. Mech. Dyn. Astron. **78**(1/4), 75–82 (2000)
12. D.G. Currie, S. Dell'Agnello, G.O. Delle Monache, A lunar laser ranging retroreflector array for the 21st century. Acta Astronaut. **68**(7–8), 667–680 (2011)
13. D.G. Currie, S. Dell'Agnello, G.O. Delle Monache, Lunar laser ranging retroreflector for the 21st century, in *17th International Workshop on Laser Ranging, Proceedings of the Conference, 16–20 May 2011, Bad Kotzing, Germany* (2011). Published online at http://cddis.gsfc.nasa.gov/lw17
14. D.G. Currie, S. Dell'Agnello, G.O. Delle Monache, K. Zacny, B. Behr, Current status and expected performance of the lunar laser ranging retroreflector for the 21st century, in *63rd International Astronautical Congress, 1–5 October 2012, Naples, Italy* (2012)
15. S. Dell'Agnello, G.O. Delle Monache, D.G. Currie et al., Creation of the new industry-standard space test of laser retroreflectors for the GNSS and LAGEOS. Adv. Space Res. **47**(5), 822–842 (2011)
16. S. Dell'Agnello, G.O. Delle Monache, D.G. Currie et al., ETRUSCO-2: an ASI-INFN project of development and SCF-Test of GNSS Retroreflector Arrays (GRA) for Galileo and the GPS-3, in *17th International Workshop on Laser Ranging, Proceedings of the Conference, 16–20 May, 2011, Bad Kotzing, Germany* (2011). Published online at http://cddis.gsfc.nasa.gov/lw17
17. S. Dell'Agnello, M. Maiello, D.G. Currie, Probing general relativity and new physics with lunar laser ranging. Nucl. Instrum. Methods Phys. Res. A **692**, 275–279 (2012)
18. S.D. Goodrow, T.W. Murphy, Effects of thermal gradients on total internal reflection corner cubes. Appl. Opt. **51**(36), 8793 (2012)
19. Google Lunar X Prize (GLXP), http://www.googlelunarxprize.org/
20. HoneyBee Corporation, http://www.honeybeerobotics.com/
21. International Space Exploration Coordination Group (ISECG), Global Exploration Roadmap (2011). http://www.nasa.gov/pdf/591066main_GER_2011_for_release.pdf
22. W. Jin, J. Li, Determination of some physical parameters of the moon with lunar laser ranging data. Earth Moon Planets **73**(3), 259–265 (1996)
23. W.S. Kiefer, P.J. McGovern, J.C. Andrews-Hanna, J.W. Head III., J.G. Williams, M.T. Zuber, the GRAIL Science Team, GRAIL gravity observations of lunar volcanic complexes, abstract #2030, in *Lunar and Planetary Science Conference, XLIV, The Woodlands, TX, March 18–22* (2013)
24. A. Khan, K. Mosegaard, J.G. Williams et al., The core of the Moon—molten or solid? in *36th Annual Lunar and Planetary Science Conference, March 14–18, 2005, League City, Texas* (2004). Abstract no. 1122
25. A. Lue, G.D. Starkman, Squeezing MOND into a Cosmological Scenario (2003), eprint arXiv: astro-ph/0310005
26. F. Lyard, F. Lefevre, T. Letellier, O. Francis, Modelling the global ocean tides: modern insights from FES2004. Ocean Dyn. **56**(5–6), 394–415 (2006)
27. M. Martini, S. Dell'Agnello, D. Currie, MoonLIGHT: a USA-Italy lunar laser ranging retroreflector array for the 21st century. Planet. Space Sci. **74**(1), 276–282 (2012)
28. D.D. McCarthy, G. Petit, 2004 IERS Conventions (2003)
29. T.W. Murphy, E.G. Adelberger, J.B.R. Battat, L.N. Carey, C.D. Hoyle, P. Leblanc, E.L. Michelsen, K. Nordtvedt, A.E. Orin, J.D. Strasburg, C.W. Stubbs, H.E. Swanson, E. Williams, The Apache Point Observatory lunar laser-ranging operation: instrument description and first detections. Publ. Astron. Soc. Pac. **120**, 20 (2008)
30. T.W. Murphy Jr., E.G. Adelberger, J.B.R. Battat et al., APOLLO: millimeter lunar laser ranging. Class. Quantum Gravity **29**(18), 184005 (2012)
31. NLSI NASA Lunar Science Institute, http://lunarscience.nasa.gov/
32. J.R. Pritchard, A. Loeb, 21 cm cosmology in the 21st century. Rep. Prog. Phys. **75** (2012). http://iopscience.iop.org/0034-4885/75/8/086901

33. N. Rambaux, J.G. Williams, The Moon's physical librations and determination of their free modes. Celest. Mech. Dyn. Astron. **109**, 85–100 (2011). Online version including tables Oct. 26, 2010, doi:10.1007/s10569-010-9314-2

34. B.G. Bills, R.D. Ray, Lunar orbital evolution: a synthesis of recent results. Geophys. Res. Lett. **26**(19), 3045–3048 (1999) (GeoRL Homepage)

35. R.D. Ray, D.E. Cartwright, Times of peak astronomical tides. Geophys. J. Int. **168**(3), 999–1004 (2007)

36. S. Sasaki, Accuracy assessment of lunar topography models. Earth Planets Space **63**, 15–23 (2011). Special Issue: New Results of Lunar Science with KAGUYA (SELENE)

37. SELENE-2, http://www.jspec.jaxa.jp/e/activity/selene2.html

38. E.C. Silverberg, D.G. Currie, Performance of the laser-ranging system at McDonald Observatory. J. Opt. Soc. Am. **61**, 692–693 (1971)

39. E.M. Standish, J.G. Williams, Orbital ephemerides of the Sun, Moon, and planets, in *Explanatory Supplement to the Astronomical Almanac*, ed. by S. Urban, P.K. Seidelmann, US Naval Observatory, Washington, DC, 3rd edn. (University Science Books, Mill Valley, 2012), pp. 305–345, Chap. 8, http://iau-comm4.jpl.nasa.gov/XSChap8.pdf, ISBN 978-1-891389-85-6

40. D.E. Smith, M.T. Zuber, G.A. Neumann, F.G. Lemoine, E. Mazarico, M.H. Torrence, J.F. McGarry, D.D. Rowlands, J.W. Head III., T.H. Duxbury, O. Aharonson, P.G. Lucey, M.S. Robinson, O.S. Barnouin, J.F. Cavanaugh, X. Sun, P. Liiva, D. Mao, J.C. Smith, A.E. Bartels, Initial observations from the Lunar Orbiter Laser Altimeter (LOLA). Geophys. Res. Lett. **37**, L18204 (2010)

41. T.K. Varghese, W.M. Decker, H.A. Crooks, Matera Laser Ranging Observatory (MLRO): an overview, in *NASA Goddard Space Flight Center, Eighth International Workshop on Laser Ranging Instrumentation* (1993), 5 p. (SEE N94-15552 03-19)

42. R.C. Weber, P.-Y. Lin, E.J. Garnero et al., Seismic detection of the lunar core. Science **331**, 309–312 (2011)

43. M.A. Wieczorek, G.A. Neumann, F. Nimmo, W.S. Kiefer, G.J. Taylor, R.J. Phillips, S.C. Solomon, J.C. Andrews-Hanna, S.W. Asmar, A.S. Konopliv, F.G. Lemoine, D.E. Smith, M.M. Watkins, J.G. Williams, M.T. Zuber, The crust of the Moon as seen by GRAIL. Science **339**, 671–675 (2013)

44. C.M. Will, K. Nordtvedt, Conservation laws and preferred frames in relativistic gravity 1: preferred-frame theories and an extended PPN formalism. Astrophys. J. **177**(3), 757 (1972)

45. J.G. Williams, D.H. Boggs, C.F. Yoder, J.T. Ratcliff, J.O. Dickey, Lunar rotational dissipation in solid body and molten core. J. Geophys. Res., Planets **106**, 27933–27968 (2001)

46. J.G. Williams, S.G. Turyshev, D.H. Boggs et al., Lunar laser ranging science: gravitational physics and lunar interior and geodesy. Adv. Space Res. **37**, 67–71 (2006)

47. J.G. Williams, A.S. Konopliv, D.H. Boggs, R.S. Park, D.-N. Yuan, F.G. Lemoine, S.J. Goossens, E. Mazarico, F. Nimmo, R.C. Weber, S.W. Asmar, H.J. Melosh, G.A. Neumann, R.J. Phillips, D.E. Smith, S.C. Solomon, M.M. Watkins, M.A. Wieczorek, M.T. Zuber, J.C. Andrews-Hanna, J.W. Head, W.S. Kiefer, I. Isamu, P.J. McGovern, C.W. Stubbs, G.J. Taylor, Lunar interior properties from the GRAIL mission. 44th Lunar and Planetary Science Conference, held March 18–22, 2013 in The Woodlands, TX. LPI Contribution No. 1719, p. 3092

48. J.G. Williams, D.H. Boggs, W.M. Folkner, DE430 Lunar Orbit, Physical Librations, and Surface Coordinates, JPL IOM 335-JW, DB, WF (2013, in preparation)

49. C.F. Yoder, Venus' free obliquity. Icarus **117**, 250–286 (1995)

50. M.T. Zuber, D.E. Smith, M.M. Watkins, S.W. Asmar, A.S. Konopliv, F.G. Lemoine, H.J. Melosh, G.A. Neumann, F. Nimmo, R.J. Phillips, S.C. Solomon, M.A. Wieczorek, J.G. Williams, S.J. Goossens, G. Kruizinga, E. Mazarico, R.S. Park, D.-N. Yuan, Gravity field of the Moon from the Gravity Recovery and Interior Laboratory (GRAIL) mission. Science **339**, 668–671 (2013). doi:10.1126/Science.1231507

51. K. Zacny, D. Currie, G. Paulsen, T. Szwarc, P. Chu, Development and testing of the pneumatic lunar drill for the emplacement of the corner cube reflector on the moon. Planet. Space Sci. **71**, 131–141 (2012). 2012. doi:10.1016/j.pss.2012.07.025

Earthquake Prediction Research Using Radio Tomography of the Ionosphere

Vyacheslav Kunitsyn, Elena Andreeva, Ivan Nesterov, Artem Padokhin, Dmitrii Gribkov, and Douglas A. Rekenthaler

Abstract Under development since its invention in 1990 as an ancillary application of ionospheric radio-tomography (RT), a new earthquake (EQ) prediction system is being evaluated. It has already been deployed along the United States West Coast, from Vancouver in Canada to San Diego in Southern California, and is currently undergoing Beta testing. This Chapter addresses RT–EQ prediction concepts, the underlying RT theory, evolution and implementation, and a few examples of the Beta test system's performance. This work is an investigation of EQ precursors, which we hope will lead to an operational system. The current system provides a foundation and the tools to study ionospheric effects linked to conditions in the Earth's crust prior to major earthquakes. Progress toward a fully operational system will require several more years of data acquisition and analysis.

List of Acronyms and Abbreviations

LORT Low Orbital (satellite in low orbit) Radio Tomography
HORT High Orbital Radio Tomography, using GNSS: GPS, Beidou, Glonass, Galileo Satellites
HO   High Orbital
LO   Low Orbital
RT   Radio Tomography
RTI   Radio Tomography of the Ionosphere
RO   Radio Occultation
AGW   Acoustic Gravity Wave
SLE   System of Linear Equations
TEC   Total Electron Count (or Content)
TID   Traveling (or Transient) Ionospheric Disturbance

V. Kunitsyn (✉) · E. Andreeva · I. Nesterov · A. Padokhin
Faculty of Physics, M. Lomonosov Moscow State University, Moscow, Russia
e-mail: kunitsyn77@mail.ru

D. Gribkov · D.A. Rekenthaler
Radio-Hydro-Physics LLC, 3400 Jennings Chapel Road, Woodbine, MD 21797, USA

F. Freund, S. Langhoff (eds.), *Universe of Scales: From Nanotechnology to Cosmology*,
Springer Proceedings in Physics 150, DOI 10.1007/978-3-319-02207-9_15,
© Springer International Publishing Switzerland 2014

UHF   Ultra-high Frequency
VHF   Very-high Frequency
HF   High Frequency

1 Introduction

The possibility of using ionospheric perturbations to recognize processes, which occur at the Earth's surface and inside the crust prior to major earthquakes, had long captured Mino Freund's imagination and had become a motivation for his far-reaching plan for flying "smart" nanosatellite constellations. This led to years of fruitful discussions with Douglas Rekenthaler and his group who are at the forefront of high-resolution ionospheric tomography.

An ionospheric RT system is a distributed sounding system, which exploits space-borne satellite beacons and a dispersed array of ground station receivers. The ground stations continuously receive the signals from the satellite beacons sounding the ionosphere along constantly changing ray paths, collecting data, which allow modeling of the spatio-temporal state of the ionospheric structure. Satellite RT provides a foundational methodology for regional, eventual global monitoring of the near-Earth ionospheric plasma, and of the total electron content (TEC) along the ray paths.

Studies of the ionosphere and the physics of ionospheric processes rely on the knowledge of the spatial distribution of the ionospheric plasma as a function of time. Because the ionosphere is the propagation medium for radio waves, the ionosphere significantly affects the performance of navigation, geo-location, reconnaissance, remote sensing, and telecommunication systems, inter alia. Beacon rays transiting the ionosphere interact with the electrons in the ionospheric plasma. They experience Faraday rotation, scintillation, scattering, refraction, and many other effects. "Noise" sources develop from the overhead magnetosphere, solar and related ionospheric disturbances, and other solar system and cosmic influencing factors. From below, lithospheric and anthropogenic effects are mapped onto the ionosphere, including the effects from rocket launches, hurricanes and other atmospheric processes, and large entity events such as earthquakes.

Therefore, investigations into the structure of the ionosphere are of interest for many practical applications. Existing satellite navigation systems hosting beacons, with a growing number of arrays of corresponding ground receiving networks, are suitable for sounding the ionosphere along different directions. The data are processed by tomographic methods, i.e. by reconstructing the temporal and spatial distribution of the ionospheric electron density.

Three principal methods of satellite ionospheric radio tomography have been successfully developed, with continuing advances on-going [1–8]. Since the early 1990s, the optimal RT systems have been based on data from LO navigation beacon satellites—LEO satellites operating at 1,000 km altitudes, in circular, near-polar orbits. In recent years, RT studies have progressed to use measurements using HO navigation systems—the Global Navigation Satellite Systems (GNSS), currently

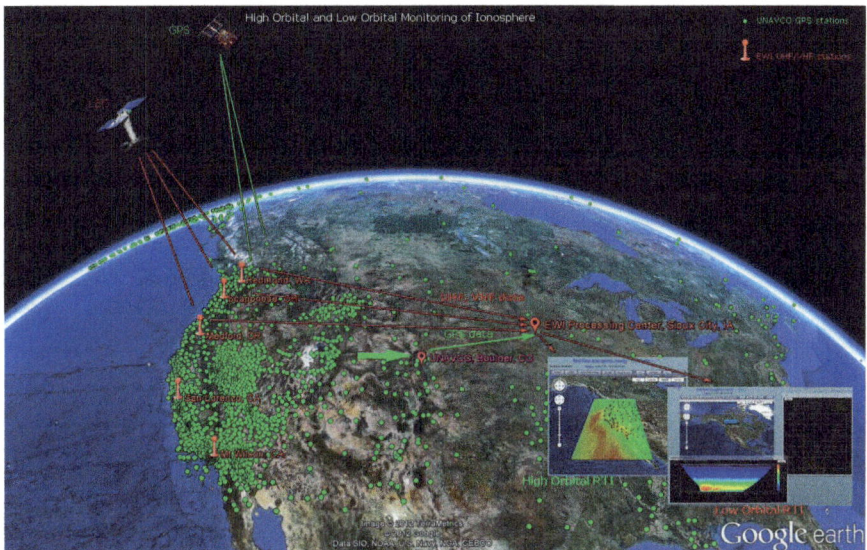

Fig. 1 LORT and HORT illustrated

comprising the US Global Positioning System (GPS), and the Russian Glonass systems which are currently operational, as well as the European Galileo and Chinese BeiDou systems which are being built. These two principal types of radio tomography are referred to as low-orbital RT (LORT) and high-orbital RT (HORT) [6–8]. A third technique currently in development is Radio Occultation (RO)—a technique of quasi-tangential sounding. The LORT and HORT techniques are depicted in Fig. 1.

The apparent motion of the HO satellites is far slower than the motion of the LO satellites; consequently, sampling of the ionosphere is faster with LO, and the resulting TEC measurements are more useful. At scales of meters to tens and hundreds of meters, the ionosphere changes in a matter of minutes, not hours, and rapid sampling is important. Figure 1 presents the scheme of satellite radio probing of the near-Earth's environment that includes the atmosphere, the ionosphere, and the protonosphere. The transmitters onboard the LO and HO satellites and the ground receivers provide the sets of rays transecting the Earth's near-space environment, and allows determination of the group and phase paths of the radio signals (in the case of LO systems, only the phase paths) along the corresponding rays. The receivers onboard the LO satellites that receive the radio transmissions from the HO satellites are also suitable for determining the group and phase paths of the signals along the set of the rays quasi-tangential to the Earth's surface. These measurements are suitable for sounding the near-space environment along various directions and calculating the integrals (or the differences of the integrals) of the refraction index in the medium. This set of the integrals can be inverted by the RT procedure for the parameters of the medium. In the case of ionospheric sounding, the integrals of the refraction index are reduced to the integrals of the ionospheric electron density.

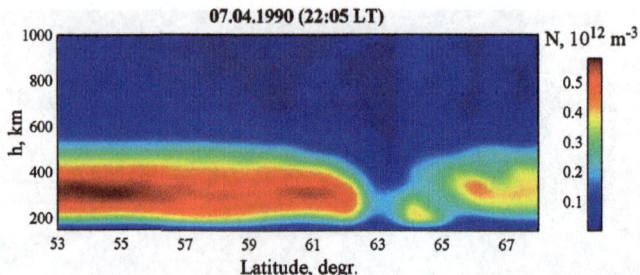

Fig. 2 LORT image of the ionosphere (Moscow-Murmansk) on April 7, 1990 at 22:05 LT

The early (circa 1990) LO navigational systems (US Navy Navigation Satellite System "Transit" and Russian Tsikada/Parus satellite system) allow the receivers to determine the differences in the Total Electron Content (TEC), typically measured in units of "electrons per cubic meter," over large regions of the earth, wherever an appropriate array of ground station receivers exists, but not continuously. The time gap between successive measurements depends on the number of operational satellites in the global constellation or satellite inventory, currently fewer than ten satellites. With over 50 satellites in high orbit, the "newer" HO systems (GNSS: GPS/GLONASS) are suitable for continuous worldwide TEC measurements.

Over the past two decades, as a result of numerous field experiments and archival data analyses, our team has confirmed the existence of unique, EQ-precursor-related signatures in the ionosphere, which are correlated in space, time and magnitude to earthquake preparation processes, i.e. earthquake precursors. These signatures have sometimes been evident tens of hours, several days, and possibly weeks, prior to the actual EQ event, and they are evident after the event for some period of time as well. The underlying cause of these signatures has not been determined; however, radon and hydrogen emanations, piston-type vertical displacement of the Earth surface, stress-induced charge flow in rock strata near fault lines, massive air ionization at ground level, and other theories have been offered. In any case, the ionospheric signatures have repeatedly demonstrated tell-tale signatures. Regardless of the primary source of the disturbances, the EQ preparation process results in ionospheric signatures which can be detected using RT methods.

2 Early Experiments with LORT

The world's first LORT images were reconstructed in March-April 1990 by geo-physicists at Moscow State University and the Polar Geophysical Institute of the Russian Academy of Sciences. One of the first RT cross sections of the ionosphere (ca. 1990) between Moscow and Murmansk is shown in Fig. 2.

The horizontal axis in this plot shows the latitude and the vertical axis shows the altitude. The ionospheric electron density is given in the units of 10^{12} m^{-3}. This image clearly shows an ionospheric trough at about 63°–65°N and a local extremum within it. Further, numerous experiments revealed the complex and diverse structure and dynamics of the ionospheric trough. In 1992, preliminary results using RT imaging of the ionosphere were obtained by colleagues from the UK [9]. These early LORT-based studies and applications drew significant interest from geophysicists around the world.

Today, more than ten research teams in different countries are engaged in these investigations. A dozen LORT receiving networks (chains of ground stations, typically dispersed along a longitudinal line at 150 km intervals) are currently operational in different regions of the world: in Russia, the US West Coast, Alaska, Great Britain, Scandinavia, Finland, Greenland, Japan, and the Caribbean region, all extensively used for research and scientific studies. A new LORT system has been built in India, and the LORT system in Southeast Asia is being upgraded. The Russian transcontinental LORT chain includes nine receiving sites arrayed along the Sochi-Moscow-Svalbard line. This is the world's longest LORT chain (about 4000 km in length), and is unique in the fact that its measurements cover a huge ionospheric sector stretching from the polar cap and auroral region to low latitudes. As a result, measurements along this chain are suitable for studying disturbance transfers between the auroral, subauroral, and low-latitude ionosphere, and for investigating the structure of the ionospheric plasma in different latitudinal regions as a function of solar, geophysical, and seasonal conditions. The Russian RT array and a corresponding RT array in Alaska are located diametrically on opposite sides of the Earth with a 12-hour time shift between them. A series of LORT experiments carried out in Europe, America, and Southeast Asia over the last twenty years has demonstrated the utility and high efficiency of RT methods for the study of diverse ionospheric structures.

3 Low Orbital Ionospheric Radio Tomography: Methodology

LORT relies on source signals from low orbiting satellites flying in nearly circular orbits at an altitude of about 1000–1150 km. The dual frequency signals from the satellite beacons are acquired by specialized ground-based receivers. The phase difference between two coherent signals transmitted from these satellites at nominal frequencies of 150 and 400 MHz are recorded and processed on the ground. The receivers are arranged in an array generally parallel to the ground projection of the satellite paths, and the distance between the neighboring receivers is typically a few hundred kilometers. The reduced phases φ recorded at the receiving sites are the input data for the RT imaging. The integrals of the electron density N along the rays linking the ground receivers with the onboard satellite transmitter are proportional to the absolute (total) phase Φ [1, 2], which includes the unknown initial phase φ_0:

$$\alpha \lambda r_e \int N d\sigma = \Phi = \varphi_0 + \varphi \qquad (1)$$

Here, λ is the wavelength of the satellite radio signal, $d\sigma$ is the element of the ray, and r_e is the classical electron radius. The scaling coefficient α (of the order of unity) depends on the sounding frequencies used. Equation (1) can be recast in the operator form [4] that includes the typical uncorrelated measurement noise ξ:

$$PN = \Phi + \xi \qquad (2)$$

where P is the projection operator mapping the two-dimensional (2D) distribution N to the set of one-dimensional (1D) projections Φ. Thus, the problem of tomographic inversion is reduced to the solution of the linear integral equations (2) for the electron concentration N.

One of the most appropriate ways to solve (2) is to discretize (approximate) the projection operator P. This yields the corresponding system of linear equations (SLE) with the discrete operator L:

$$LN = \Phi + \xi + E, \qquad E = LN - PN \qquad (3)$$

where E is the approximation error that depends on the solution N itself. Note that Eqs. (2) and (3) are equivalent if the approximation error E is known. However, in the case of reconstructing the data of a real RT experiment, E is not known, and, in fact, quite a different SLE is actually solved:

$$LN = \Phi + \xi \qquad (4)$$

SLE (4) is not equivalent to SLE (3). In other words, the difference between the solutions of (3) and (4) ensues from the difference in both the quasi-noise component and the correlated (in time and along the rays) approximation error E.

For SLE (4) to be solved, the absolute phase Φ together with φ_0 should be known. The errors in φ_0 estimated by the different receivers can result in contradictory and inconsistent data, which leads to low-quality RT reconstructions. In order to avoid this difficulty, a method of phase-difference radio tomography (RT based on the difference of the linear integrals along the neighboring rays) was developed [10], which does not require the initial phase φ_0 to be determined. The SLE of the phase-difference RT is determined by the corresponding difference:

$$AN = LN - L'N = \Phi - \Phi' = D + \xi \qquad (5)$$

where $LN = \Phi$ is the initial SLE and $L'N = \Phi'$ is the system of linear equations along the set of neighboring rays.

There are numerous algorithms, both direct and iterative, to solve SLEs (4) and (5). At present, for problems of ray radio tomography of the ionosphere, iterative algorithms are most popular; however, non-iterative algorithms are also used. These algorithms utilize a singular value decomposition with its modifications, regularization of the root mean square (RMS) deviation, orthogonal decomposition, maximum entropy, quadratic programming, and Bayesian approaches, inter alia. Extensive numerical modeling and LORT imaging of numerous experimental data reveal efficient combinations of the various methods and algorithms which yield the best reconstructions.

"Phase-difference" LORT provides much better results and higher sensitivity compared to traditional phase measurement methods. This is confirmed by reconstructions of experimental data. The horizontal and vertical resolution of LORT in

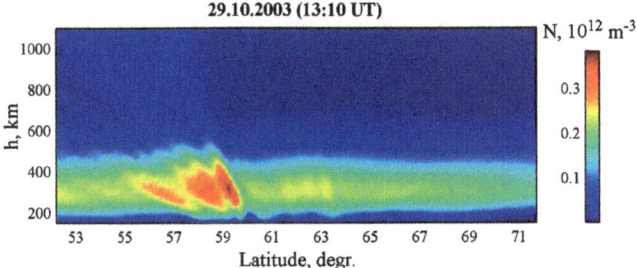

Fig. 3 LORT image of the ionosphere above the Alaska region on October 29, 2003 at 13:10 UT

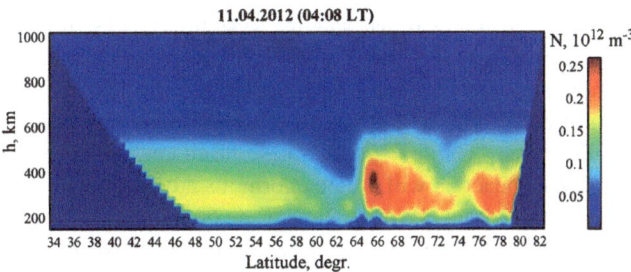

Fig. 4 LORT Image of the ionosphere (Moscow-Svalbard) on April 11, 2012 at 04:08 LT

its linear formulation is 20–30 km and 30–40 km, respectively. If the refraction of the rays is taken into account, the spatial resolution of LORT can be improved to 10–20 km.

LORT reconstructions reveal a series of the irregularities such as various waves, wavelike structures, and other features. Figure 3 presents an example of a complex wave-like perturbation with a distinctive wave-front, which was observed during the Halloween storm of 2003 in the Alaska region.

It is worth noting that the ionospheric plasma can have a highly complex structure even in undisturbed conditions, as illustrated by Fig. 4, which shows the LORT cross section of the ionosphere between Sochi and Svalbard during geomagnetically quiet conditions ($Kp < 1$).

Here, wavelike disturbances with a characteristic size of 50 km are seen above Svalbard (78°–79°N). In the central segment of the image (59°–65°N) the electron density decreases. In the southern part of the cross-section (42°–55°N), wavelike structures with a spatial period of 100–150 km are apparent. A wide ionization trough in the interval of 62°–64°N is observed on the LORT reconstruction. The local maximum at 65°–66°N is almost merged with the polar wall of the trough. A spot of enhanced ionization is identified within the trough about 63°–64°N latitude. Additionally, wavelike disturbances are revealed throughout the area bounded by 66°–78°N.

Thus, LORT is capable of reconstructing nearly instantaneous 2D snapshots of the electron density distribution in the ionosphere, which typically cover a time span of 5–15 minutes. The time interval between successive RT reconstructions, dependending on the number of operational satellites is now about 30–120 minutes. The LORT method is also suitable for determining the plasma flows by analyzing the successive RT cross sections of the ionosphere [11]. If the LORT receiving segment consists of several receiving arrays located within a few hundred km of each other, that configuration will allow 3D imaging of the ionosphere. The necessity to have multiple receiving chains is the major limitation of LORT.

4 Recent HORT-based Investigations

The evolution of global navigational satellite systems (GNSS, including the US Global Positioning System, GPS, and the Russian GLONASS) opened new opportunities to continuously measure trans-ionospheric radio signals and to attack the inverse problem of radio sounding. The current GNSS inventory will soon be augmented by the European "Galileo" and Chinese "BeiDou" satellite systems. Collectively, these GNSS beacons will create enormous advantages for RT programs. Today, GNSS beacon signals are continuously recorded at a large number of regional and global receiving networks. For example, the network operated by the International GNSS Service, IGS comprises about two thousand receivers. These data are suitable for reconstruction of the ionospheric electron density.

The inverse problems of radio sounding based on GNSS data, which pertains to the general method of tomographic characterization with incomplete data, are inherently high-dimensional. Due to the relatively low apparent angular velocity of the high-orbiting GNSS satellites, allowance for the temporal variations of the ionosphere becomes essential. This makes the RT problem four-dimensional (three spatial coordinates, plus time) and exacerbates the incompleteness of the data: every point in space is not necessarily traversed by the rays that link the satellites and the ground-based receivers; therefore, data gaps arise in the regions when only a few receivers are available. The solution of this problem requires special mathematical approaches.

The method for HORT ionospheric sounding typically exploits the phases of the radio signals that propagate from the satellite to the ground receiver at two coherent multiple frequencies. For example, in GPS-based soundings, these frequencies are $f_1 = 1575.42$ MHz and $f_2 = 1227.60$ MHz. The corresponding data (L_1 and L_2) are the phase paths of the radio signals measured in units of wavelengths of the sounding signals. Another parameter that can be used in the analysis is the pseudo-ranges (the group paths of the signals)—the time taken by the wave-trains at the frequencies f_1 and f_2 to propagate from the satellite source to the ground-based receiver. The phase delays L_1 and L_2 are proportional to the total electron content, TEC, the integral of electron density along the ray between the satellite and the receiver:

$$\text{TEC} = \left(\frac{L_1}{f_1} - \frac{L_2}{f_2} \right) \frac{f_1^2 f_2^2}{f_1^2 - f_2^2} \frac{c}{K} + const, \tag{6}$$

where $K = 40.308 \text{ m}^3 \text{ s}^{-2}$ and $c = 3 \times 10^8$ m/s is the speed of light in vacuum. Note that, by using the phase delay data, it is only possible to calculate the TEC value up to a certain constant indicated as the additive term in formula (6). The relationship (6) is similar to formula (1) with the unknown constant in the right-hand side of the system.

The TEC values can also be derived from the pseudo-ranges P_1 and P_2 [12]:

$$\text{TEC} = \frac{P_2 - P_1}{K(\frac{1}{f_2^2} - \frac{1}{f_1^2})} \tag{7}$$

However, compared to the phase data, the pseudo-range data are strongly distorted and contaminated by noise. The noise level in P_1 and P_2 is typically 20–30 % and even higher, while in phase data it is below 1 % and rarely reaches a few percent. Therefore, for HORT, the phase data are preferable.

Most authors solve the HORT problem using a set of linear integrals. In that approach, it is assumed that the TEC data are sufficiently accurately determined from the phase and group delay data. However, the absolute TEC (7) is determined with a large uncertainty in contrast to the TEC differences that are calculated highly accurately. Therefore, the phase difference approach was applied in this case as well. In other words, instead of the absolute TEC, the corresponding differences or the time derivatives $d\text{TEC}/dt$ were used as the input data for the RT problem.

The problem of the 4-D GNSS-based radio tomography can be solved by the approach developed in 2-D LORT. In this approach, the electron density distribution is represented in terms of a series expansion of certain local basis functions; in this case, the set of linear integrals or their differences is transformed into SLE. However, in contrast to 2-D LORT, here it is necessary to introduce an additional procedure, which interpolates the obtained solutions in the area with missing data. The implementation of this approach in the regions covered by dense receiving networks (e.g., North America, Japan, Alaska, and Europe) with a rather coarse calculation grid and suitable splines of varying smoothness has proved to be highly efficient.

Another approach seeks sufficiently smooth solutions of the problem so that the algorithms provide a good interpolation in the area of missing data. For example, let us consider a Sobolev norm and seek a solution which minimizes this norm over the infinite set of solutions of the initial (underdetermined) tomographic problem (5):

$$AN = D, \quad \min_{AN=D} \|f - f_0\|_{W_n^2} \tag{8}$$

Here the function f is the solution with a given weight.

Practical implementation of this approach faces difficulties associated with the solving the constrained minimization problem. The direct approach utilizing the method of the Lagrange's undetermined multipliers results in SLE with high-dimensional (due to the great number of the rays) matrices, which do not possess any special structure which would allow one to simplify the solution. Therefore, one approach is to solve this minimization problem by an iterative method which is

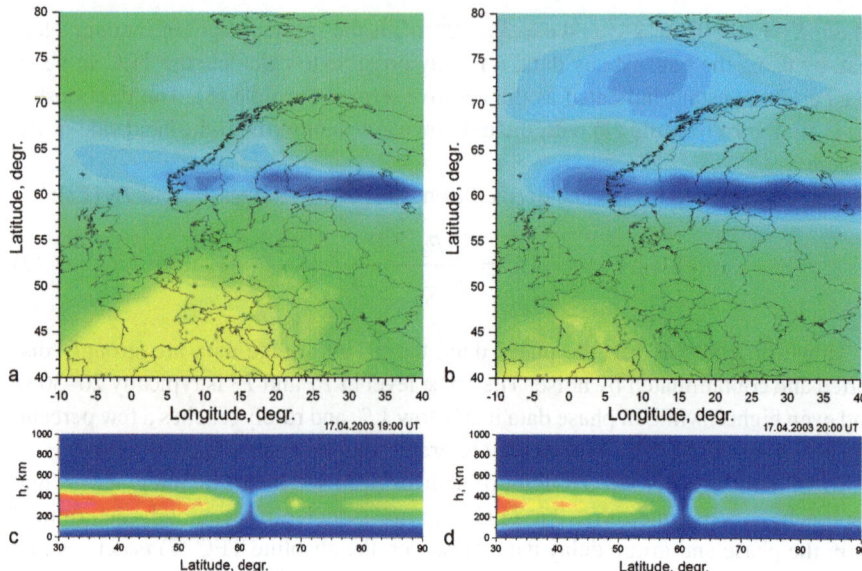

Fig. 5 Ionospheric HORT reconstructions over Europe on April 17, 2003: (**a, c**) 19:00 UT, (**b, d**) 20:00 UT. (**a, b**): TEC maps in the latitude-longitude coordinates; the color scale is from 0 to 35 TECU (1 TECU $= 10^{16}$ m^{-2}). (**c, d**): Meridional cross sections along 21°E in the latitude-altitude coordinates; the color scale is from 0 to 0.6×10^{12} m^{-3}

a version of the SIRT technique, with additional smoothing (by filtering) of the iterative increments over the spatial variables. This method allows *a priori* information to be introduced through both the initial approximation for iterations, and through weighting coefficients that determine the relative intensity of the electron density variations at different heights.

Computer-aided modeling shows that quasi-stationary structures are reconstructed with a reasonable accuracy although HORT has a significantly lower resolution than LORT. As a rule, the vertical and horizontal resolution of HORT is 100 km at best, and the time step (the interval between two consecutive reconstructions) is currently 20–60 minutes. In regions covered by dense receiving networks (Europe, North America, Japan, and Alaska), the resolution can be improved to 30–50 km with a 30–10 min interval between the consecutive reconstructions. However, a resolution of 10–30 km, with a time step of 2 minutes can now be achieved in the few regions which have very dense receiving networks—California and Japan. Obviously, there exists the potential for even near-real-time updates as the densification of networks improves in selected areas: the cost of these ground station receivers is relatively low.

The reconstructions presented below illustrate the possibilities of the HORT techniques as depicted in Fig. 8. Figure 5 displays the evolution of the ionospheric trough above Europe in the evening on April 17, 2003. The TEC maps and the meridional cross sections along 21°E show the trough widening against the background overall nighttime decrease in electron density.

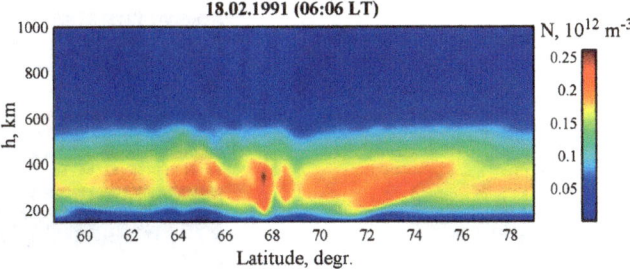

Fig. 6 Wavelike structures formed in the ionosphere 30 minutes after launch of a rocket from the Plesetsk Cosmodrome

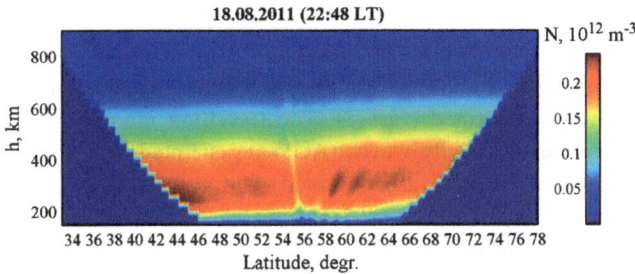

Fig. 7 Typical ionospheric cross-section through the heated area of the ionosphere

5 Ancillary Investigations

Besides being suitable for reconstructing large-scale ionospheric phenomena of natural origin, LORT is also efficient for tracking artificial ionospheric disturbances as illustrated in Fig. 8. The LORT cross-section in Fig. 6 shows the wavelike structures formed in the ionosphere 30 minutes after launch of a rocket from the Plesetsk Cosmodrome. The cosmodrome is located approximately 63°N 200 km distant from the satellite sub-track. Such anthropogenic disturbances have quite a complex structure wherein large irregularities (200–400 km) coexist with smaller ionospheric features (50–70 km). The slope of the "wavefront" is also varying.

Wave disturbances generated by launching high-power rocket vehicles are described in [13] where it is shown that the start of the rocket creates acoustic-gravity waves (AGW), which induce corresponding perturbations in the ionospheric TEC. During RT experiments with the Moscow-Murmansk array, very long-lived local disturbances in the ionospheric plasma were identified above sites where ground-based industrial explosions were carried out [14].

RT methods have revealed generation of ionospheric disturbances by the Sura heating facility (see Fig. 7), which radiates high-power radio-frequency (high frequency, HF) waves modulated with a 10-min RF on/off heating period [15].

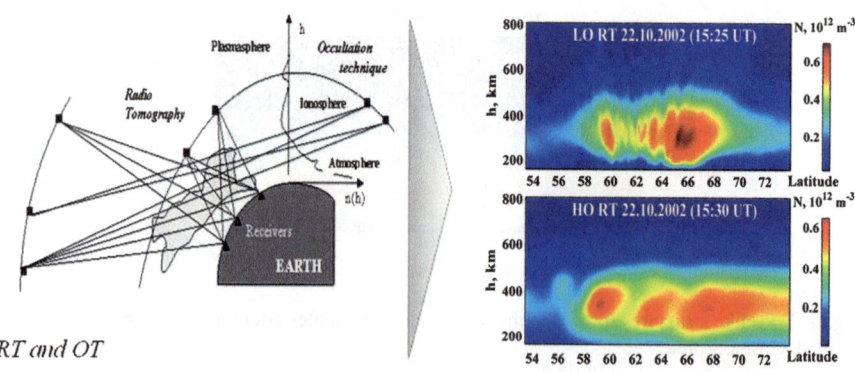

Fig. 8 Occultation technique in context with HORT and LORT, the two primary RT methodologies

A narrow trough in the ionization, aligned with the propagation direction of the heating HF wave, is identified (Fig. 7). The travelling ionospheric disturbances associated with acoustic gravity waves (AGWs) generated by the Sura heater diverging from the heated area, are observed. Unfortunately, insufficient density of the HO receivers in central Russia prevented us from obtaining high-quality HORT images of the ionosphere during this heating experiment; however, the data recorded by a few available receivers readily support the presence of the AGWs [15].

6 RT Compared to Alternative Ionospheric Sounding Techniques

There are a small number of existing low-earth-orbiting satellites—FormoSat-3/COSMIC and a few others—which have the capability of recording GNSS signals. Those satellites exemplify the radio occultation technique (OT), in that they capture quasi-tangential measurements of the electron density N [16–18]. The OT method—essentially reception of GNSS signals on LO satellites—results in sounding the ionosphere across a wide range of different geometries of the transmitting and receiving systems. The OT method, with integrals of N over a set of quasi-tangential rays (the satellite-receiver links) is a particular case of the RT method; therefore, it will be necessary to eventually integrate the occultation data into the general RT scenario [7, 8, 19]. The combination of RT and OT, where the RT data are supplemented by the satellite-to-satellite sounding (OT) data, would noticeably improve vertical resolution of RT reconstructions.

Existing ultra-violet (UV)-sounding systems (GUVI, SSULI, FormoSat-3/COSMIC) provide integrals of N squared, and the UV sounding data can be incorporated into a general tomographic iterative program. For example, as the first step of an ionospheric reconstruction, an iterative approach using linear integrals and RT data alone is used. Then, based on the distribution of electron density N obtained with the first iteration, we could run iterations for N squared with the UV data. Subsequently, we transform the distribution of N squared derived at this step of the reconstruction into the distribution of N (the result of the second iteration); this distribution can be further used in a third iteration. Odd iterations will only work with the radio sounding data, while even iterations will use the UV input. Overall, we obtain the tomographic products which use both the radio sounding and UV sounding data. However, in order to ensure convergence and to obtain high-quality final results, experimental data of different kinds should be consistent and have commensurate accuracy; otherwise, additional iterations based on the "bad" data will only degrade the result. Unfortunately, as of today, UV data are far less accurate that the data of navigation radio sounding.

We note that the described RT methods refer to the ray tomography [1] that neglects the diffraction effects. In our previous work, we developed the methods for diffraction tomography and statistical tomography [1, 2, 4, 7]. The diffraction tomography is applicable for imaging the structure of isolated localized irregularities with allowance for the diffraction effects. The statistical tomography reconstructs the spatial distributions of the statistical parameters of the randomly irregular ionosphere [7, 20].

In numerous experiments, the RT images of the ionosphere were compared with the corresponding parameters (vertical profiles of electron density and critical frequencies) measured by ionosondes [4, 7, 21–27]. In most cases, the RT results closely agree with the ionosonde data within the accuracy of both methods.

These features indicate that strong spatial gradients in electron density typical in the region of equatorial anomaly can cause the discrepancies in the plasma frequencies calculated from RT and from the ionosonde measurements.

In experiments on vertical pulsed sounding of the ionosphere, the signal is not reflected from directly overhead. Even in the case of vertical sounding of horizontally stratified ionosphere, the ordinary wave tends to deviate toward the pole, and in the point of reflection, it becomes perpendicular to the local geomagnetic field [28]. Therefore, in the general case, the reflection does not occur vertically above the sounding point but somewhat away from overhead.

It is worth noting that the ionosonde measurements during the geomagnetically disturbed periods are often unstable because the ionosphere experiences significant transformations on short time scales that alter the radio propagation conditions. In particular, the electron density N in the D-region ionosphere sharply increases, and, due to strong radio absorption, most ionograms do not show any reflections. In contrast to the ionosondes, which are essentially HF radars, RT methods are suitable for imaging the ionosphere even during the strongly disturbed solar and geophysical conditions, because the high sounding frequency used in the RT applications (150 MHz) allows one to neglect the absorption.

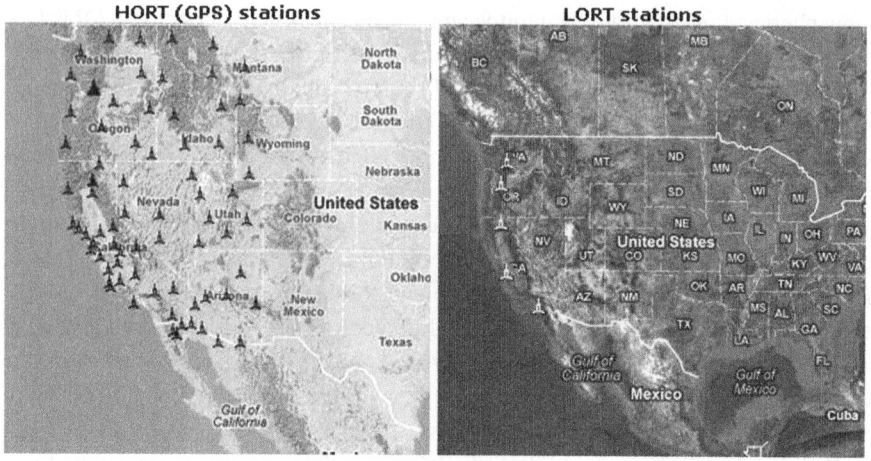

Fig. 9 US West Coast prediction system, currently using 100+ existing GPS receiving stations, and 5 LORT ground station receivers

7 New US West Coast Earthquake Prediction Network: Operational and Successful

During an early RT experiment, conducted between Beijing and Manila, a magnitude 7+ earthquake occurred, and evidence of the earthquake was readily apparent as a unique anomaly on the ionospheric maps which our Moscow State University team generated from that test. On closer inspection, it was apparent that those signatures were not only evident at the time of the earthquake: they actually existed several days prior to the earthquake—as precursors. Moreover, the signatures were unlike any others—no "ambient or normal" ionospheric signatures theretofore observed matched those seen prior to the Manila event. A new EQ prediction methodology was evident.

In the intervening years this methodology has been tested in many different locations, with observations of precursors prior to scores of events. Those have been recorded using GPS/HORT, and UHF/VHF LORT signals. Although there were no ground stations with data communications for pre-EQ alerting, using after-the-fact data, it has been possible to show strong precursors using these ionospheric "reconstruction" techniques for large Sumatra and Chile earthquakes, as two examples. The providence of the RT-based EQ prediction method evolved from the Moscow State University team in Moscow, ca. 1990. These methods have since been further developed and automated, not only for earthquake prediction, but also including tsunami detection and early warning.

With regard to an EQ forecasting system, it is important that the system demonstrate reproducible results and be available 365/24/7. It should be a reliable system for imaging the ionosphere (for convenience we use the terms imaging, reconstruction, or characterization). An EQ prediction system should have low probability of

Fig. 10 Chilean EQ, Images of the TEC obtained using a very limited number of GPS ground stations (represented by *dots* in the image). The signals from the GPS ground stations were processed to display the TEC changes in the ionosphere. The eventual location of the magnitude 8.8, Chilean earthquake is at the apex of the blue peak-shape on the West Coast of Chile. These GPS-TEC images were obtained a day prior to the 2010 Chile EQ M8.8

Fig. 11 A precursor two days prior to the Chilean EQ. This precursor illustrates a signature which is "lower" than the normal ambient background level

Fig. 12 The ionosphere depicted immediately after the EQ, showing a "hot spot" over the epicenter

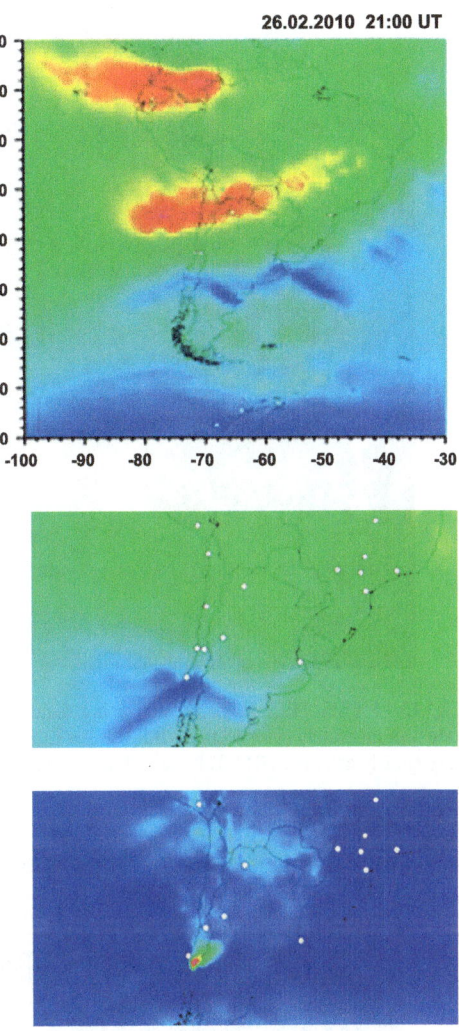

false alarms, and offer a high probability of successful predictions in terms of time-liness, magnitude, and geo-position of the resulting EQ. The new US West Coast system shown in Fig. 9 is a hybrid assembly, which synthesizes both GPS/Glonass HORT with UHF/VHF LORT. There are only a few LORT satellites in orbit to-day. We have worked with cubesat and nanosat designers to arrive at concepts for affordable new satellites in a robust constellation, to fill this need in the future.

The system uses all available GPS ground station receivers—typically as many as 340. The timeliness of the data from those receivers is a limitation, as is the reliability of the receivers themselves. Most of the HORT system downtime experienced over the past two years of Beta operation has resulted from data transmission failures at the receiving sites. The system also uses 5 UHF/VHF ground station receivers for the LORT data. Those are a variant of the NWRE ITS33 receivers,

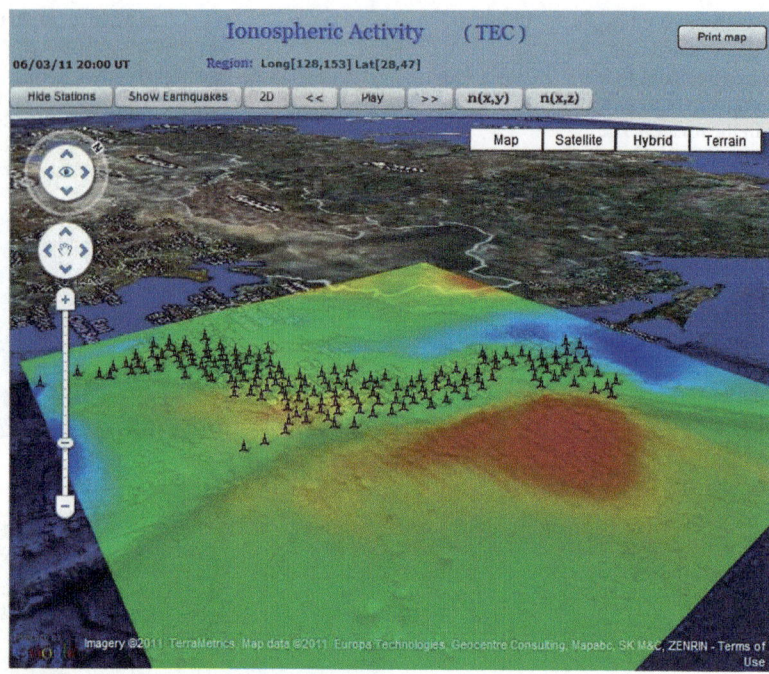

Fig. 13 A GPS-TEC image created using data from 250 GPS ground stations in Japan (small icons) immediately prior to the 2011 Tohoku EQ. This ionospheric reconstruction was completed very shortly after the 2011 Tohoku EQ, using archived data

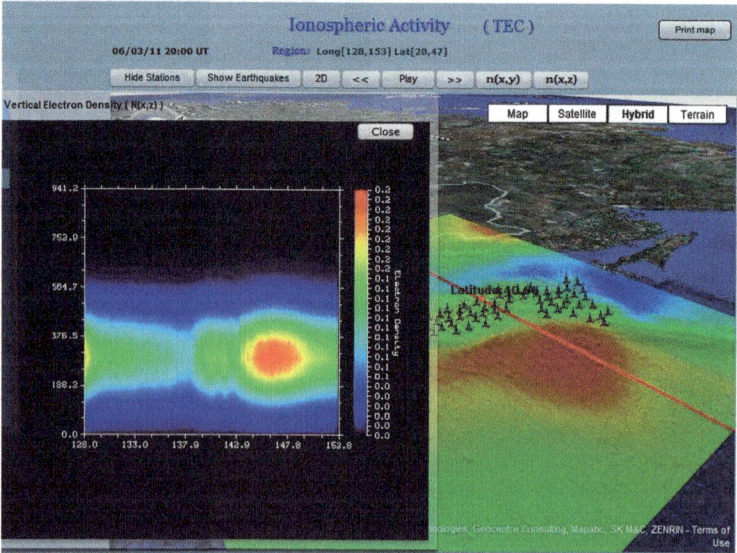

Fig. 14 Another view of the Tohoku EQ, as reconstructed using HORT data. Ionospheric signatures were apparent approximately 5 days prior to the EQ

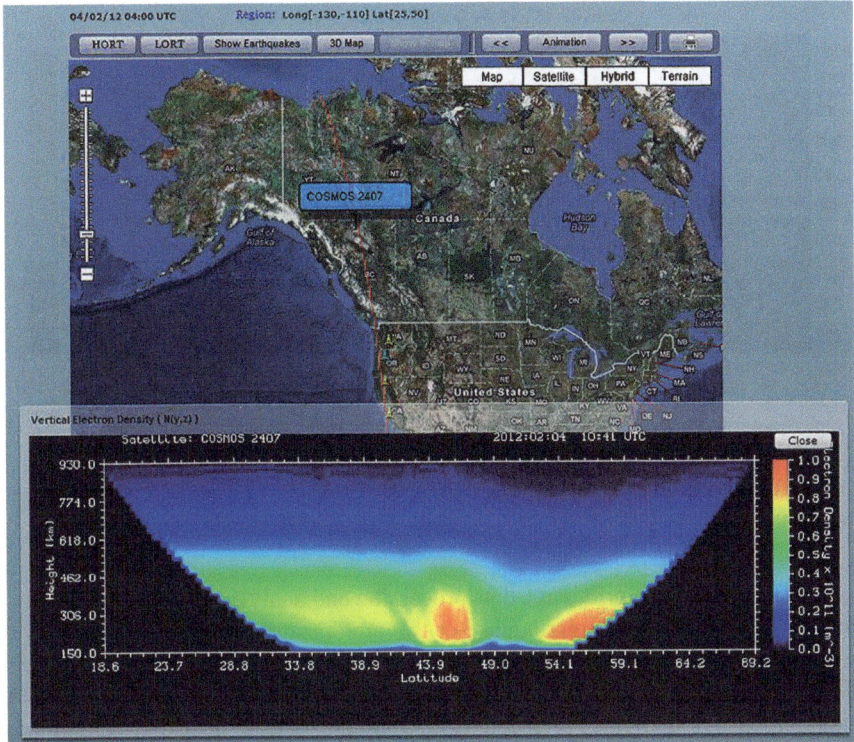

Fig. 15 M5.7 Earthquake: Vancouver Island 2012-02-04 20:05:32; 48.867N, 127.875W: Changes in the ionosphere appeared ∼9 hours before the earthquake, showing the ground track of the Cosmos 2407 satellite, source of the UHF/VHF beacon signals used for the ionospheric reconstruction. This was a LORT technique

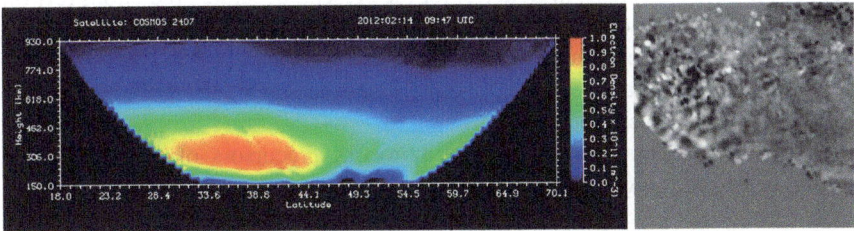

Fig. 16 An M6 Earthquake off the coast of Oregon, 2012-02-15 03:31:20; 43.536N, 127.380W: Changes in the ionosphere appeared 17 hours prior to the EQ, and the first acoustic gravity waves near the epicenter become visible in the vertical 2D-projection on the right ∼1 hour before the EQ

with added timing boards and improved broadband oscillators specifically designed for this application. The ground stations are designed to seek, acquire, and transmit dual frequency, phase and amplitude beacon signals from the satellites. Doppler corrections are automatically accounted for. The satellite data is then merged with

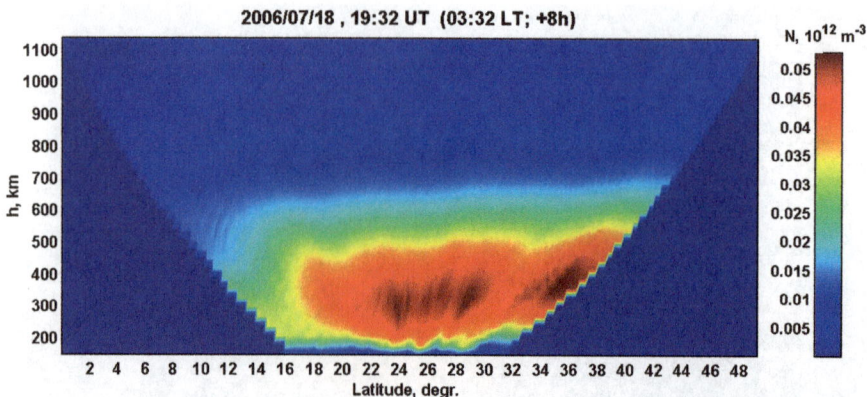

Fig. 17 This, and the following three figures depict precursors at various time-early prior to EQs occurring near Taiwan. These are excellent examples of the wave and TEC structures characteristic of EQs in this region. EQs 07/23–27 ($M = 4.3$–4.4); EQ 07/28, 07:40UT ($M = 5.9$) Lat = 24.18, Long = 122.53

satellite position ephemerides at a data processing and analysis center in Sioux City, Iowa, safely outside of any earthquake or tsunami hazard zones. From Sioux City, the processed data are further transmitted to Wisconsin and Illinois, where various proprietary routines are used to make the actual predictions. The entire system is automated, autonomously returning to operational status in the event of lightning strikes, power outages, or other anomalies which might otherwise hinder operations. To-date, there have been six successful predictions by the US West Coast system, each ranging from a few hours to three days prior to the EQ event. There have been two missed events: one due to system downtime for upgrade of receiver software, and one due to a failure of the data source. Those predictions have not been advertised, as the system remains in Beta test for the present: accurate predictions rely on an extensive database of ambient background signals, and on experience with the unique attributes of the precursory EQ signatures in a particular geographic area, as well as time of day, season, external influences on the signals, *inter alia*. Prediction accuracy improves over time.

Ionospheric precursors clearly exist; however, precursors of different earthquakes, in different regions, do not necessarily exhibit a common pattern. For example, earthquakes may be preceded by either increases or decreases in electron density, or by specific inhomogeneities, or waves, or turbulence patterns, *inter alia*. Whichever features the ionospheric precursors may be in a given case, our system is suitable for 2-D or 3-D imaging of the ionosphere, and for distinguishing between the precursors and the ionospheric structures associated with the usual solar-geophysical perturbations. The means used in previous studies—ionosondes, Global Ionospheric Maps (GIM) technology, and similar) were incapable of unambiguously differentiating between these factors.

An illustration of the technique is shown in the HORT-only images (see Figs. 10–20).

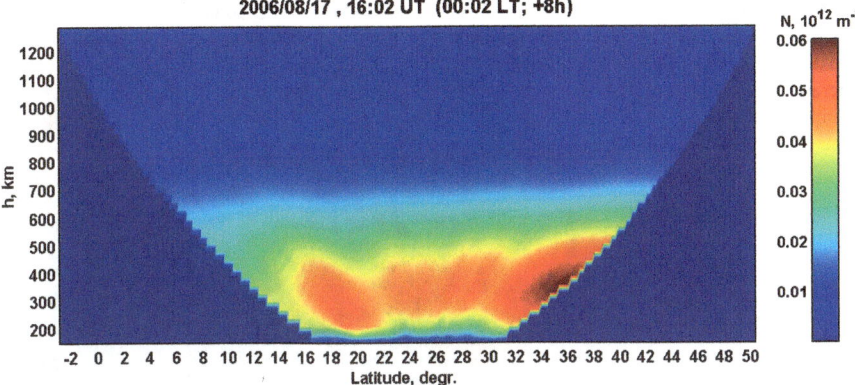

Fig. 18 Taiwan EQ Precursor. EQs 08/19–25 ($M = 4.0$–4.6); EQ 08/27, 17:11UT ($M = 5.4$) Lat $= 24.95$, Long $= 122.94$

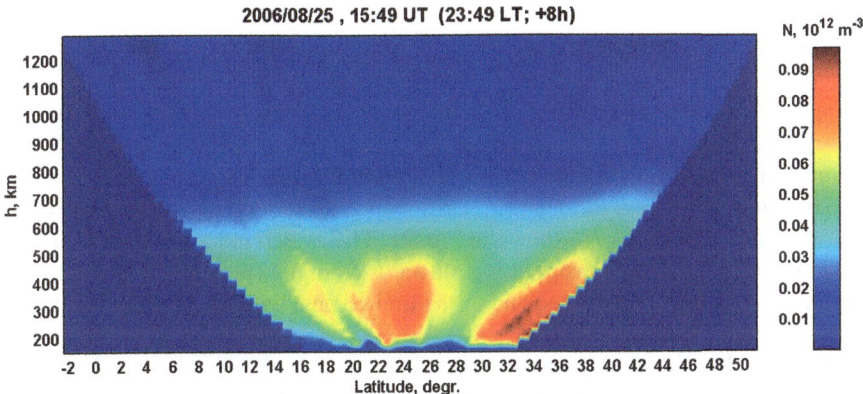

Fig. 19 Taiwan EQ Precursor. EQ 08/27, 17:11UT ($M = 5.4$) Lat $= 24.95$, Long $= 122.94$

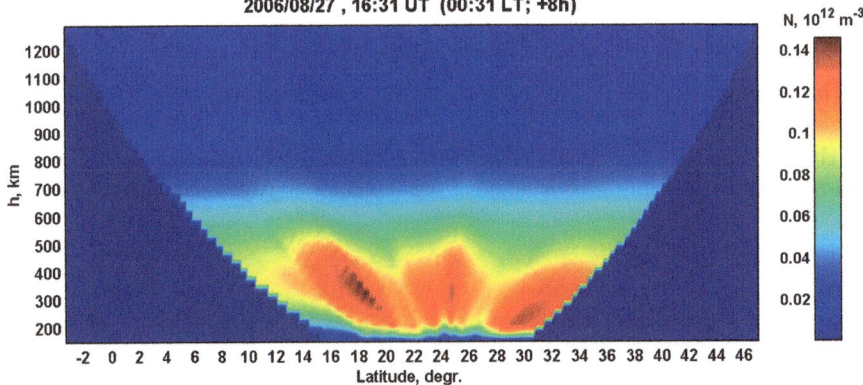

Fig. 20 Taiwan EQ Precursor. EQ 08/27, 17:11UT ($M = 5.4$) Lat $= 24.95$, Long $= 122.94$

Fig. 21 The diverging disturbance caused by the acoustic gravity waves generated by the Tohoku earthquake

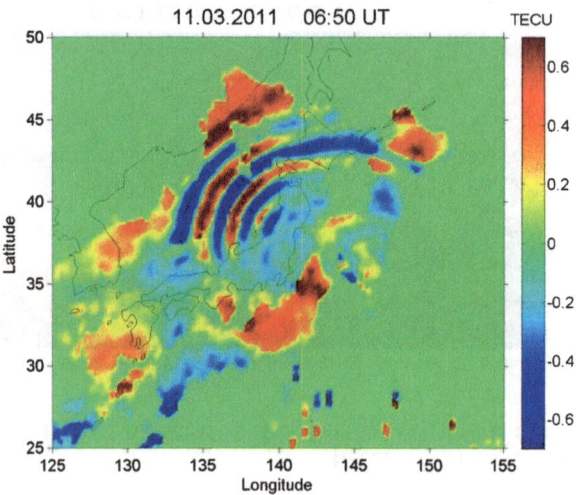

8 Tsunami Early Warning

Using archived data, we analyzed the ionospheric disturbances over and near Japan after the magnitude 9.2 Tohoku earthquake. A sub-set of data from the 1200 GPS ground station receivers in Japan were analyzed by HORT methods, with a very high time resolution (2–3 min) in [29]. Disturbances observed in the vertical TEC an hour after the main shock are shown in Fig. 21. TEC waves induced by earthquake-generated AGWs are seen propagating outwards from the epicentral area. The spatial limits of the diagram correspond to the limited area within which the receiving network is sufficiently dense.

As shown by the measurements at Hawaii (Figs. 22 and 23) and on the US West Coast (Fig. 24), the ionospheric acoustic-gravity waves (AGW) appear well ahead of the tsunami ocean waves, leading it by 1–2 hours at the US West Coast. This allows us to predict the tsunami wave a long time in advance of the arrival of the ocean waves, and to estimate its parameters from the ionospheric data. Numerical modeling of the atmospheric and ionospheric effects of the tsunami crest propagation confirms the aforementioned conclusions concerning formation of the accompanying ionospheric AGW, and accumulation of the frontal waves in the course of the tsunami wave propagation.

The RT methodology has the potential to "predict" earthquakes, and to "detect" and provide early warning of tsunamis. The development of the system continues apace.

9 Conclusion

Ionospheric precursors clearly exist; however, precursors of different earthquakes, in different regions, do not necessarily exhibit a common pattern. For example, the

GPS-derived TEC data from Hawaiian receivers and the corresponding wavelet

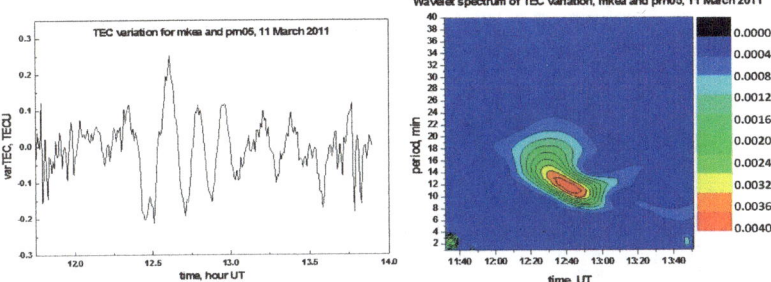

- Hawaii is situated at approximately two thirds of the distance from Japan to the next densest GPS networks in Southern California. Here, wave-like structures are seen in the ionosphere approximately one hour ahead of the tsunami ocean waves. The arrival of the tsunami at DART buoy 51407, located near the Big Island of Hawaii (19.60 N. 203.50 E), was predicted at 13:17 UT.

Fig. 22 GNSS based (mkea IGS station) reconstruction of the Tohoku-induced ionospheric waves over Hawaii. Wave-like structures are seen in the ionosphere approximately one hour before the tsunami ocean waves struck the Islands (according to DART buoy 51407 (19.6N, 203.5E) at 13:17UT)

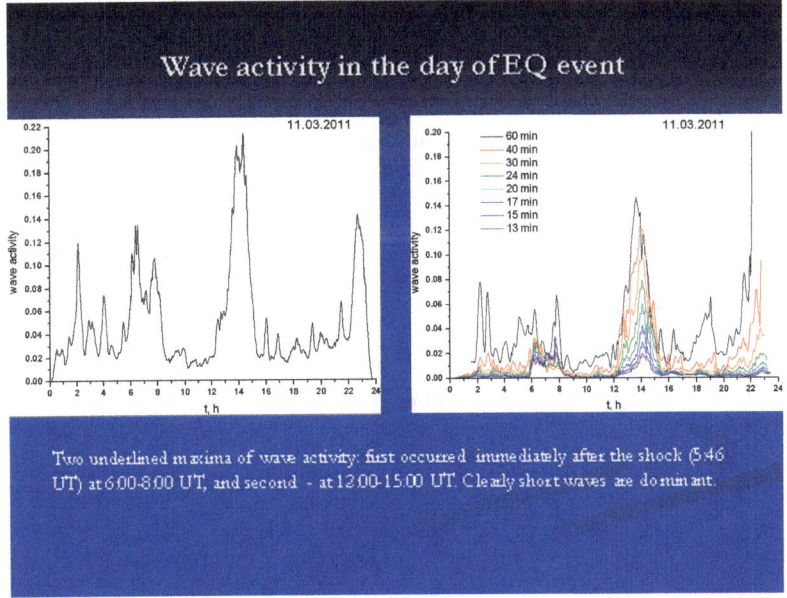

Fig. 23 Figure showing HORT wave activity on the day of the Tohoku EQ. That same event resulted in ionospheric waves over Hawaii a few hours later, but significantly before the actual tsunami water waves hit the islands. Still later, as depicted in Fig. 24, the ionosphere showed perturbations over the US West Coast—waves which were detected by our US West Coast HORT and LORT system

Far-zone ionospheric disturbances
(using GPS networks in Southern California)

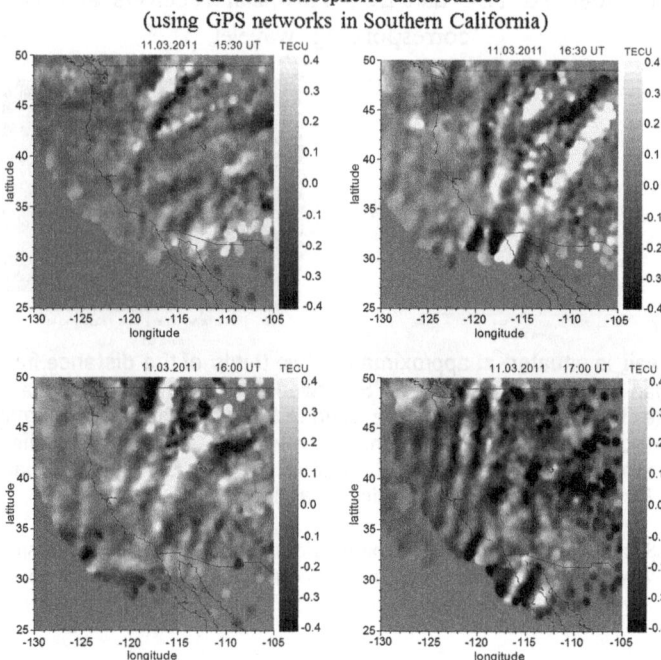

Fig. 24 Tohoku-induced ionospheric waves over the US West Coast, as they appeared hours before the ocean waves hit the coastal areas

earthquakes may be preceded by either increases or decreases in electron density, or by specific inhomogeneities, or waves, or turbulence patterns, inter alia. Whichever features the ionospheric precursors may be in a given case, our system is suitable for 2-D or 3-D imaging of the ionosphere, and for distinguishing between the precursors and the ionospheric structures associated with the usual solar-geophysical perturbations. The means used in previous studies—ionosondes, Global Ionospheric Maps (GIM) technology, inter alia—were incapable of unambiguously differentiating between these factors. The RT methodology has the potential to "Predict" earthquakes, and to "Detect" and provide early warning of tsunamis. The development of the system continues apace.

References

1. V.E. Kunitsyn, E.D. Tereshchenko, *Tomography of the Ionosphere* (Nauka, Moscow, 1991) (in Russian)
2. V.E. Kunitsyn, E.D. Tereschenko, Radiotomography of the ionosphere. IEEE Antennas Propag. Mag. **34**, 22–32 (1992)
3. R. Leitinger, Ionospheric tomography, in *Review of Radio Science 1996–1999*, ed. by R. Stone (Science Publications, Oxford, 1999), pp. 581–623

4. V.E. Kunitsyn, E.D. Tereshchenko, *Ionospheric Tomography* (Springer, Berlin, 2003)
5. S.E. Pryse, Radio tomography: a new experimental technique. Surv. Geophys. **24**, 1–38 (2003)
6. G.S. Bust, C.N. Mitchell, History, current state, and future directions of ionospheric imaging. Rev. Geophys. **46**, RG1003 (2008). doi:10.1029/2006RG000212
7. V.E. Kunitsyn, E.D. Tereshchenko, E.S. Andreeva, *Radio Tomography of the Ionosphere* (Nauka, Moscow, 2007) (in Russian)
8. V.E. Kunitsyn, E.D. Tereshchenko, E.S. Andreeva, I.A. Nesterov, Satellite radio probing and radio tomography of the ionosphere. Usp. Fiz. Nauk **180**(5), 548–553 (2010)
9. S. Pryse, L. Kersley, A preliminary experimental test of ionospheric tomography. J. Atmos. Terr. Phys. **54**, 1007–1012 (1992)
10. E.S. Andreeva, V.E. Kunitsyn, E.D. Tereshchenko, Phase difference radiotomography of the ionosphere. Ann. Geophys. **10**, 849–855 (1992)
11. V.E. Kunitsyn, E.S. Andreeva, S.J. Franke, K.C. Yeh, Tomographic investigations of temporal variations of the ionospheric electron density and the implied fluxes. Geophys. Res. Lett. **30**(16), 1851 (2003). doi:10.1029/2003GL016908
12. B. Hofmann-Wellenhof, H. Lichtenegger, J. Collins, *Global Positioning System: Theory and Practice* (Springer, Berlin, 1992)
13. R. Ahmadov, V. Kunitsyn, Simulation of generation and propagation of acoustic gravity waves in the atmosphere during a rocket flight. Int. J. Geomagn. Aeron. **5**(2), 1–12 (2004). doi:10.1029/2004GI000064
14. E.S. Andreeva, M.B. Gokhberg, V.E. Kunitsyn, E.D. Tereshchenko, B.Z. Khudukon, S.L. Shalimov, Radiotomographical detection of ionosphere disturbances caused by ground explosions. Cosm. Res. **39**(1), 13–17 (2001)
15. V.E. Kunitsyn, E.S. Andreeva, V.L. Frolov, G.P. Komrakov, M.O. Nazarenko, A.M. Padokhin, Sounding of HF heating-induced artificial ionospheric disturbances by navigational satellite radio transmissions. Radio Sci. **47**, RS0L15 (2012). doi:10.1029/2011RS004957
16. G. Hajj, R. Ibanez-Meier, E. Kursinski, L. Romans, Imaging the ionosphere with the global positioning system. Int. J. Imaging Syst. Technol. **5**(2), 174–187 (1994)
17. E. Kursinski, G. Hajj, W. Beritger et al., Initial results of radio occultation of Earth atmosphere using GPS. Science **271**(5252), 1107–1110 (1996)
18. Y.A. Liou, A.G. Pavelyev, S.S. Matyugov et al., *Radio Occultation Method for Remote Sensing of the Atmosphere and Ionosphere* (InTech, Rijeka, 2010), Ed. Y.A. Liou
19. E.S. Andreeva, N.A. Berbeneva, V.E. Kunitsyn, Radio tomography using quasi tangential radiosounding on traces satellite-satellite. Geomagn. Aeron. **39**(6), 109–114 (1999)
20. E.D. Tereschenko, M.O. Kozlova, V.E. Kunitsyn, E.S. Andreeva, Statistical tomography of subkilometer irregularities in the high-latitude ionosphere. Radio Sci. **39**, RS1S35 (2004). doi:10.1029/2002RS002829
21. L. Kersley, J. Heaton, S. Pryse, T. Raymund, Experimental ionospheric tomography with ionosonde input and EISCAT verification. Ann. Geophys. **11**, 1064–1074 (1993)
22. V.E. Kunitsyn, E.S. Andreeva, O.G. Razinkov, E.D. Tereschhenko, Phase and phase-difference ionospheric radiotomography. Int. J. Imaging Syst. Technol. **5**(2), 128–140 (1994)
23. J. Heaton, S. Pryse, L. Kersley, Improved background representation, ionosonde input and independent verification in experimental ionospheric tomography. Ann. Geophys. **13**, 1297–1302 (1995)
24. J. Heaton, G. Jones, L. Kersley, Toward ionospheric tomography in Antarctica: First steps and comparison with dynasonde observations. Antarct. Sci. **8**, 297–302 (1996)
25. J. Heaton, P. Cannon, N. Rogers, C. Mitchell, L. Kersley, Validation of electron density profiles derived from oblique ionograms over the United Kingdom. Radio Sci. **36**, 1149–1156 (2001)
26. R. Dabas, L. Kersley, Radio tomographic imaging as an aid to modeling of ionospheric electron density. Radio Sci. **38**(3), 1035 (2003). doi:10.1029/2001RS002514

27. S.J. Franke, K.C. Yeh, E.S. Andreeva, V.E. Kunitsyn, A study of the equatorial anomaly ionosphere using tomographic images. Radio Sci. **38**(1), 1011 (2003). doi:10.1029/2002RS 002657
28. K.C. Yeh, C.H. Liu, *Theory of Ionospheric Waves* (Academic Press, New York, 1972)
29. V. Kunitsyn, I. Nesterov, S. Shalimov, Japan Earthquake on March 11, 2011: GPS–TEC evidence for ionospheric disturbances. JETP Lett. **94**(8), 616–620 (2011)

Pre-Earthquake Signals at the Ground Level

Jorge A. Heraud P.

Abstract Can we really predict earthquakes? Will we be able to do it sometime? The answer to the first question is no, we still cannot predict earthquakes but we seem to be moving in the right direction. At the ground level, several electromagnetic manifestations previous to rupture, are slowly fitting into place.

The main theme in this chapter is luminescence and the prevailing hypothesis in the case described here is the electric origin of the phenomenon commonly known as *EarthQuake Lights*, or EQLs. The difficulty of dealing with luminescence nowadays is separating any EQLs from noise arising from artificial lights, electric short circuits, sparks, even fire from electric power lines, from substations, circuit breakers and the like. The San Lorenzo Island off the coast of Lima, Peru has provided three very outstanding cases of pre-seismic and co-seismic EQLs—with geological consistency—spanning 266 years of observation, including two high magnitude earthquakes. In addition there are three cases linked to low magnitude events with close-by hypocenters, about 2 km, which produces pre-earthquake EQLs on the island. In these cases, the high stress resulting from the build-up of a large magnitude earthquake produced a 21-day anticipation of the EQLs, whereas the low magnitude earthquake gave rise to a short 38 hours lead time. New picture evidence collected at San Lorenzo island show rock formations at an old colonial times prison reported to have been the focus of luminescence evidence before the mega earthquake in 1746. In 2007, a strong M8.0 earthquake 160 km away from the island, produced co-seismic lights, probably via the local activation of positive hole carriers by passing seismic waves, specifically S waves, igneous rocks forming vertical dykes in the bay of Lima.

Videos taken by security cameras on the PUCP campus show a very good time correlation with ground acceleration records from a seismometer located on the campus. Videos from an off-campus location show lights that were generated at a hill at the southern end of the city and were confirmed by qualified eyewitnesses. Observations from the San Lorenzo Island point to the possibility that small rocky

J.A. Heraud P. (✉)
Director, Institute for Radio Astronomy (INRAS), Pontificia Universidad Católica del Perú (PUCP), Lima, Peru
e-mail: jheraud@pucp.edu.pe

F. Freund, S. Langhoff (eds.), *Universe of Scales: From Nanotechnology to Cosmology*,   133
Springer Proceedings in Physics 150, DOI 10.1007/978-3-319-02207-9_16,
© Springer International Publishing Switzerland 2014

islets in the Bay could have been the points of origin of columns of light seemingly arising out of the ocean.

The deployment of magnetometers, in collaboration with Quakefinder, is currently building up a station network along the seismically very active Southern Peruvian coast. All in all, at least on this side of this subcontinent, luminescence seems to be coupled with the generation and transport of electric charges.

The answer to the question whether it will be able to predict earthquakes sometimes in the future is strongly linked to our ability to (i) understand the physics of rocks under stress and (ii) develop a worldwide network of ground stations to collect and process multivariate data that will allow for meaningful deductions of the data leading to predictions. This is the final quest. Wiring different types of sensors to monitor electromagnetic activity prior to earthquakes is the geophysical equivalent to an electrocardiogram except that is aimed at anticipating impending catastrophic seismic activity. Rather than just sensing the passing of mechanical waves, as cardiologists do by "feeling" the cardiac pulse at the wrist, a worldwide web of monitoring stations, combined with the Internet, might bring us early warning signs pointing at future heart attacks of mother Earth.

1 Introduction

Natural hazards mitigation has moved forward in the past years for many types of disasters, through research effort and operating measures derived from space activities, like weather satellites, increased involvement of national and international organizations and many forms of cooperation and funding. This is the case of increasingly reliable weather reports, storms, hurricane watch including path and intensity prediction, Tsunami watch, floods and volcanic activity. What about earthquakes?

One of the most important questions we are now addressing with regards to drastic seismic activity leading to destruction and loss of lives is: Can we really predict earthquakes? Will we be able to do it sometime? Many authors have looked into this issue trying to go beyond the classical research in seismicity and the statistical estimation of future fracture areas with the hope it will lead to a more physical interpretation of what happens to matter subject to great pressures, dynamic forces and other physical interactions leading to seismic rupture and earthquakes. Research could probably lead to new interpretations of the ways nature might have to communicate the advent of a future seismic event, with enough lead time so as to devise operational methods to make life-saving prediction viable.

For over one hundred years, the electrical nature of matter has been recognized through successful theories linked to experimental evidence in physics and chemistry. Astrophysics is finally completing a picture in which we might be able to identify additional ways in which nature sets in motion charges and radiation, its very constituents, to produce a broader spectrum of messengers of telluric consequences. However it is still difficult to accept that, under mechanical stress, electrons in orbitals and the lack of electrons in certain orbitals, so-called positive holes, can set them in motion. Thus, electrons and p-holes and their physical, chemical and electromagnetic interaction with matter can conceivably set up visible, measurable and recordable ways of tracing, in space and time, activity deep in the Earth's crust in anticipation of the build-up of rupture areas at the onset of an earthquake. The experimental techniques of radio-science can thus be used to study the electromagnetic consequences of the motion, acceleration and deceleration of positive and negative electric charges, the generation of electric currents in the first case and of radio frequency waves in the other two. Besides the production of local magnetic fields and their interaction with the earth's magnetic field, there is accumulation of charges and the build-up of polarization in structures and the rush of huge amounts of electric charges into narrower high points leading to corona discharges, i.e. the burst-like ionization of dielectric air to produce luminescence. In the recombination process, line emissions in the infrared are produced as well as other consequences of the drift and diffusion of charges and the variations of the height of the ionosphere. Some of these phenomena can probably propagate far enough to produce effects at or farther away from the future rupture point or hypocenter, others are probably caused locally, as the seismic wave propagates carrying energy that temporarily stresses appropriate types of rocks with higher igneous content.

In a recent paper, Friedemann Freund [1] covers this issue thoroughly, analyzing the fact that non-seismic signals occurring before earthquakes have been reported but not taken too seriously into consideration for multiple reasons. Among them, we should consider the seriousness of the reports, the uncertainty about their origin and the time correlation between the occurrence of the phenomenon and the onset of the earthquake, perhaps hours, days or weeks later. Several more could be added. The fact that most of these signals are non-linearly related to earthquake magnitude, the location of epicenters and luminous signals in uninhabited areas, the effect of deep geological structures in the propagation of electric charges, the time of the day effect in luminous phenomena for instance, constitute difficulties in their study.

The intention is to give an observational approximation, in this case bridging over centuries, to one of the most ancient records of connections between electromagnetic phenomena and seismic activity. By using an outstanding video recording of co-seismic luminescence at a strategic point in the Peruvian coast during a major earthquake in 2007, an old and curious report of luminescence in a prominent island that occurred over two hundred years ago in the same spot, will be brought into historical perspective. Even more so, in July 27, 2012, a day and a half before an earthquake, pre-seismic lights were observed again in exactly the same spot as in 1746 and 2007. Besides, the possibility of detecting pre-seismic luminescence, opens another way hitherto difficult to be turned into an operational alert method, but at least scientifically possible and desirable as a means to correlate electrically produced luminescence with other methods or prediction.

2 Seismic Luminescence: "Earthquake Lights"

For hundreds, even thousands of years, reports about the luminous phenomena previous to (pre-seismic) or coincidental (co-seismic) with earthquakes have been piling up since ancient times. Unexpected lights in the sky, rapidly flashing "flames" emanating from the ground, fire-like phenomena or "tongues of fire", globular gaseous brightness of different sizes, extended white and light blue luminescence in the sky and other shapes and forms of various colors, have all been described. Literature contains vivid descriptions of these phenomena, accounts of word-of-mouth information passed along for decades and even centuries, though most of them associated with religious beliefs related to "divine punishment".

Atmospheric events of luminous characteristics associated with earthquakes are commonly known as EQLs for "EarthQuake Lights" and have been described by Richter [2] in 1958, but observationally reported by Terada [3, 4] in 1930 and 1931. One of the first review papers on earthquake lights was written by John Derr [5] in 1973 and it gives a thorough description, with many illustrations, of luminous phenomena associated with earthquakes since one of the first analysis of the problem of considering EQLs in seismology in 1942. As will be recalled further ahead, most of the morphology described by John Derr resembles very closely the luminosities described by qualified witnesses during the M8.0 earthquake in Pisco, Peru and observed extensively in Lima, 150 km away [8]. The statistics of observations seems to be stationary over tens of years in time and thousands of kilometers in space.

In the Mediterranean area, EQLs have been reported since ancient times. Papadopoulos [6] in 1999 collected information from 30 earthquakes and concluded that EQLs were reported from rather shallow and strong earthquakes ($M \geq 6.0$) and epicenters up to about 140 km away. In Quebec, Canada, EQLs were reported by St-Laurent [7] in 2000.

2.1 Signal to Noise Ratio

One of the difficulties experimental research has to deal with leads to the enhancement of the ratio of signal to noise in any observation, collection of data and data analysis and processing. Seismic luminescence, in a figurative way, is no exception and even at present time, separating cases worthwhile looking into from hoax, is not easy. Electrical flashes associated with the power grid, shorting of power cables, oscillating high voltage transmission lines bouncing against each other, arcing in circuit breakers, corona discharges, open fires and other activities of man-made origin, usually happen. Going back into the past to study the reports of luminescence from time prior to electricity has its advantages since there is no need to mask out non-existent man-made electrical causes. However we have to deal with religious interpretations and superstitious beliefs that greatly increase the number of reports, enhance the atrocity with which the phenomena is perceived, leads to induced opinions that range from the exaggerated intensity of the phenomena to attributed relationships with other occurrences because of assumed divine purpose and control. So ancient reports of luminous phenomena may have no source of confusion with man-made causes but could be highly distorted by subjective personal and collective beliefs.

2.2 The Big Earthquake in Lima, 1746

On October 28th, 1746 at 10:30 pm (LT), a 4-minute long mega earthquake hit Lima, the capital city of the Viceroyalty of Peru, then under Spanish rule, and Callao its sea port, about 15 km due west. Without seismometers and modern seismological science, the magnitude of the earthquake cannot be known but estimates place it around 8.6. However, judging by the high percentage of the population killed and reports on the distance reached inland by the sea water of the giant Tsunami, two Spanish "leagues" or around 5 km, some seismologists have estimated a $M > 9$ event, the largest earthquake ever to strike Peru. The earthquake has been described by different sources, as a major event, much stronger than the great Lima earthquakes of 1582, July 9, 1586, October 10, 1687 or the Cusco 1650, Tacna August 13, 1868 or the Ancash, May 30, 1970, the highest human lives toll earthquake in Peru. In this event,over 75,000 people died as a consequence of the huge induced landslide of a glacier at 6,700 m above sea level that buried the city of Yungay, about 400 km north of Lima.

In 1746, Lima's population was estimated between 50,000 and 60,000, so the death toll of 7,000 was very high at over 10 %. The port of Callao was about 7,000 people and records account that about only 200 survived, about 23 ships were literally thrown through the air by the tsunami reaching midway into the city of Lima. Diego de Esquivel y Navia [9] and Perez-Mallaina [10] have written good accounts on the events previous, during and after the big earthquake of 1746 and other news from Lima newspapers in 1791, in particular *Mercurio Peruano* [11], have permitted us not only to understand the social events, the fears and disbeliefs surrounding such a tragic event, but from a more scientific point of view, to discover, buried in the noise of fear and superstition, the presence of EQLs, up to three weeks before the earthquake. More recently, Charles Walker [12] a historian steeped in Peruvian culture and history, published a book describing the social events that surrounded the tragic events related to the mega earthquake and tsunami in Lima in 1746.

2.2.1 San Lorenzo Island in the Bay of Lima

In central Peru, the bay of Lima is one of the most prominent geographical features on the Pacific coast and spans the capital city of Lima, nowadays with nine million inhabitants. San Lorenzo is the largest of a group of islands, islets and rocks protruding from the ocean, apparently with the appropriate characteristics for propagating electric charges to the surface, as will be shown. San Lorenzo is part of a group of islands together with *El Fronton*, *Palomino*, *Cabinzas* and smaller islets and rocks known as *Horadadas* and is located 4 km from the peninsula at the northern part of the bay. It is 8.3 km long, 2.3 km wide and its highest point rises to 396 m above sea level. It is arid, it has no sources of fresh water and its terrain is rocky and covered by sandy soil. Occupancy has been only temporary along centuries but pre-Columbian ruins found, give a hint that a permanent population might have been established.

Fig. 1 Old prison and
dungeons in the rock at San
Lorenzo island. EQLs in the
form of tongues of fire were
reported coming out of the
ground and walls on October
7, 1746, 21 days before the
large October 28, 1746
earthquake (Photo credit
Maria del Pilar Fortunic)

Nowadays, only a naval base exists as a permanent station but no large electric grid installations as power lines are present so as to suspect man-made electric flashes to be confused with EQLs.

2.2.2 Pre-seismic EQLs in San Lorenzo Is., October 1746

Diego de Esquivel y Navia [9], describes a particular incident in which 21 days before the earthquake, strange lights described as "tongues of fire" were seen by the captain of a sloop anchored in San Lorenzo island. They were coming out of the storeroom, the tower and fence walls, made in those days of adobe and rocks and he thought it was a product of his imagination. The old prison, shown in Fig. 1, sits on rock and in an adobe construction, so its walls could not have been the source of real fire, as will be dealt with in the following description. Some of the high security dungeons were caves and holes on the rock with a simple iron gate outside, as can be seen to the right of the building, one of them in white.

When the captain of the sloop met prisoners working at 2–3 am, he learned from the officer in charge of the prison, Manuel Romero, that he had let them out to work because he was afraid the tongues of fire would melt down the prison walls. The captain then realized that what he had witnessed earlier was not the product of his imagination but was corroborated by the prison official. The following translation, with the full account of the pre-seismic EQL event, tries to respect the description and construction of XVIII century Spanish, as contained in the following paragraph:

"On October 7th, 21 days before the sad topic of this letter, don Juan Felix Goycoechea, a man over 50 years old, native of Fuenterrabio in Guipuzcoa, Spain, captain of the king's sloop in which they were transporting rocks from San Lorenzo island to the prison, between 2 and 3 o'clock in the morning, saw the storeroom, other rooms, towers and fence walls burning; this not only worried him but filled him with horror. A little less than an hour later, he met the prisoners, subject to forced labor on that island, as they were coming to load rocks on the ship and this was a surprise for the captain who did not expect the unusual timing for this job; he asked them for the reasons for their untimely arrival and their answer was that—the island Captain, don Manuel Romero, who was very frightened, released us from prison at about 3 am, so we could watch the prison melt down and there is nobody left there who is not an eye witness of such flame and fire. With this response Captain Goycoechea, confirming what at the beginning he had taken as a false view and presumed as produced by his imagination, made public to the inhabitants of the prison the flames of fire he had seen burning, explaining the reason for the tragic forecast he had issued, saying they had been made in prevention of trouble so they could anticipate the situation with penance".

It is truly significant that this event could be rescued from the noise level of fantasy that accompanied the great earthquake in Lima in 1746, and becomes significant also in the time difference of 21 days between its occurrence and the seismic event. San Lorenzo island is a mountainous desert island that rises up to about 400m above sea level and at about 5 to 6 km off the tip, as an off shore extension of a small peninsula in the northern end of the bay of Lima. It is made mainly of sedimentary rock but part of it is made of mafic igneous rocks, which resisted erosion. As will be explained, the very intense luminous phenomena video recorded during the 2007 Pisco earthquake and seen co-seismically in Lima, away from the epicenter, occurred among other places, also precisely above San Lorenzo island.

2.2.3 Pre-seismic EQLs in San Lorenzo Is., July 2012

Mr. Diego E. Menendez, an engineer and the brother of a research assistant at our Institute, is quite familiar with our 2007 EQLs video since he helped us reformatting it. Hence he is a qualified observer for this matter. On July 27, 2012, he alerted us of outstandingly similar but brief lights he saw in the direction of San Lorenzo, the evening of July 27, 2012 at 10:30 pm LT. It was quite a welcomed coincidence but there was no hint as when the seismic event would occur, however the fact that San Lorenzo was precisely under the observed luminescence led us to think of the likelihood of a nearby event. Thirty eight hours later, on July 29, 2012 at 12:36 pm LT, a Ml 4.5 earthquake occurred just about 2 km NE of San Lorenzo Island at a depth of 58 km, corroborating the pre-seismic nature of the luminescence, reported to us ahead of the event. The scenario of these pre-seismic sightings in the bay of Lima is shown on Fig. 2. It has been estimated that the 1746 earthquake epicenter probably occurred around the same place.

2.3 The Pisco, Peru, 2007 Earthquake

Most earthquakes in Peru are caused by convergent plate tectonics, as a consequence of the subduction of the Nazca plate under the South American continental plate, one of the most active regions in the world, resulting in the rise on the Andean cordillera. According to Tavera and Bernal [13], the $Mw = 8.0$ earthquake off the coast of Pisco on August 15, 2007, was the largest shallow earthquake in Central Peru during the last 250 years. Tavera et al. [14] have covered the earthquake which produced extensive damage to property and infrastructure in the cities of Ica, Pisco, Chincha, El Carmen and many smaller towns. Total death toll was 520 people, about 1500 wounded, 59,000 houses destroyed, and Pisco was the hardest hit with 85 % of its houses destroyed.

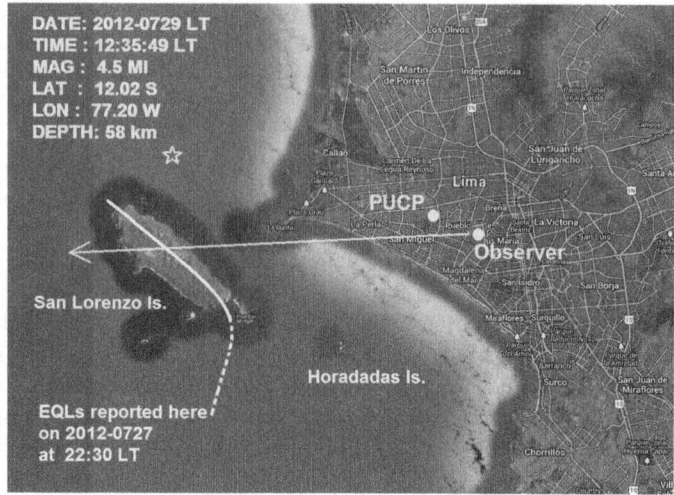

DATE: 2012-0729 LT
TIME : 12:35:49 LT
MAG : 4.5 MI
LAT : 12.02 S
LON : 77.20 W
DEPTH: 58 km
☆

San Lorenzo Is.

EQLs reported here
on 2012-0727
at 22:30 LT

PUCP
Observer

Horadadas Is.

Fig. 2 Scenario of Pre-seismic EQLs in San Lorenzo island, Lima, Peru, July 27, 2012

2.3.1 Co-seismic EQLs Along the Peruvian Coast

According to Ocola and Torres [15], co-seismic luminosity has been reported by hundreds or thousands of people along the coast, almost as far south as Nazca and as far north as Huacho, especially in small towns and beaches. Most places and small islands along the central Peruvian coast (with green captions) and referred to here, are depicted in Fig. 3, which shows the zone spanning from Nazca to Huacho. In the capital city of Lima, about 160 km north of the epicenter, thousands of people reported seeing the lights in various forms: quick, moving flashes of different colors but predominantly bright white or light blue in the direction of the coast, over the ocean, over the top of islands and some hills. Ocola describes very important evidence of lights reported by people who were on the beach at the time. They observed lights over the uninhabited islands off the coast of Chincha and hence strongly enhancing the fact that no man-made electric sources could have been responsible for the luminous phenomena. My personal experience during several interviews with eyewitnesses in two towns along the coast about 60 km south of Lima, Chilca and Pucusana, has been very similar. Observers were very emphatic in describing the luminous phenomena in the direction of the ocean and those in Pucusana (see Fig. 3) all pointed to the top of a close-by island as the origin of most EQLs.

2.3.2 Co-seismic EQLs in the Bay of Lima: Witnesses' Accounts

Complete video evidence of co-seismic luminescence was recorded at the PUCP campus by a stationary security color camera, pointing at a fixed direction with time stamp in milliseconds [16]. Using another video, recorded in a shopping center

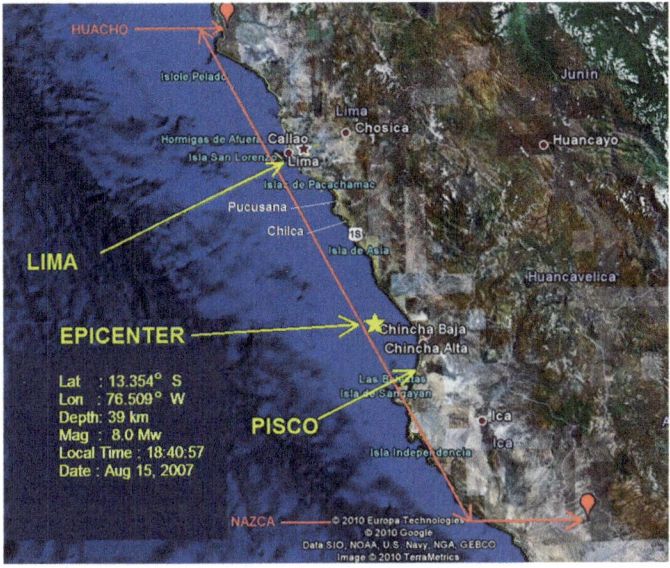

Fig. 3 Scene of reported EQLs from areas south to north of the epicenter in central Peru. Islands are shown in *green*, locations in *white*

overlooking the ocean, ray tracing could be performed to analyze the direction of the captured lights. With the additional participation of several qualified eyewitnesses, a fairly good layout of the EQLs along the Lima skyline was reconstructed.

On August 15, 2007 at 18:41:00 LT, Mr. Giancarlo Crapesi a Peruvian private pilot was landing a twin engine turbo prop plane at "Jorge Chavez" international airport in Lima. At an altitude of 1500 ft ($\sim$300 m) on his final approach, he reported to the control tower the presence of rapidly moving, flickering lights in different parts of the bay, especially on the top of islands and hills. He was told an earthquake was underway at that very moment. Mr. Capresi decided to check the status of the landing strip before committing to landing. However he decided to land instead of continuing to the only alternate airport, that of Pisco. At the time, it was not yet known that the epicenter of the $Mw = 8.0$ earthquake had actually started near Pisco so that landing there would have been impossible. In a follow-on interview Mr. Crapesi described with great detail the shapes, colors and timing of the lights, and he was present as an art student transferred his description to a Google earth map for his approval. Mr. Crapesi emphasized the lights on top of the hills along the landing path well known to him under the names "*La Regla*" and "*La Milla*". He also mentioned EQLs emanating from the top of San Lorenzo island and from "Morro Solar", a well known 350 m high hill in the southern part of Lima. From his flight experience, he described the lights as rapidly moving, similar to those produced by electrostatic discharges from wing tips of aircraft. A comprehensive image of this witness description, together with a general overview of the scenario is given later in Fig. 9.

Fig. 4 View of EQLS over San Lorenzo and El Fronton islands, and the ocean between them and the coast, as described by air traffic controller at the Lima airport

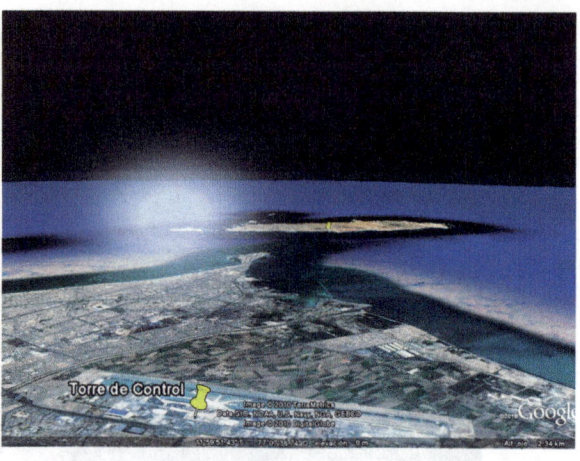

Our second eyewitness was Mr. Jorge Merino, an air traffic controller from COR-PAC, the aeronautical authority in Peru, who was on duty at the control tower at "Jorge Chavez" international airport at the time of the earthquake. Mr. Merino described the lights as "coming out of the ocean" from the strait between San Lorenzo and Fronton islands as being round in shape and somewhat diffused. His description was drawn by an art student and later reproduced on a Google earth map for his approval, as shown on Fig. 4.

The third eyewitness was the chief security officer at the Lima international airport, Mr. Juan Salas, who was near the parking ramp when the earthquake started. As he paced back to the main airport building, he saw white and light blue lights reflected off the large glass windows of the terminal. The lights were coming from the west, i.e. from the direction of the ocean. Mr. Salas saved the videos from the outside security cameras, which were compared with videos of updated cameras installed later, in order to determine the exact direction the cameras had been pointing at the time of the earthquake.

Our fourth eyewitness, second lieutenant Guillermo Zamorano from the Peruvian Navy, was at San Lorenzo Island at the time of the earthquake. He described seeing "columns of light" emanating from the ocean as he looked towards the Lima coast line. The lights were rising at four consecutive times during the earthquake, consistent with the time capture on the video at the PUCP campus. The lights were white and light blue with some spiral stripe structure. At first this witness account seemed difficult to fit since it would be difficult to understand how a column of light would rise off the ocean. However, upon further examination, a group of pointed rocks or islets was identified, low above the water line and difficult to see, but definitely above water in the precise area described by our witnesses and known as "*Horadadas*" Islands. The image, recreated with the help of our Art student, is shown in Fig. 5.

Three specific views of the southern (Morro Solar) and northern (islands) parts of the bay of Lima are shown in Fig. 6.

Fig. 5 Columns of light
emerging from little islets
(Horadadas Is.) on the ocean,
as described by an observer
on the San Lorenzo Island

"Horadadas" Islets

distance from the coast
shown : about 6.8 km

**El Fronton & San Lorenzo
islands at the northern end
of the Bay of Lima**

distance from the coast shown :
about 10 km

**El Morro Solar Hill at the
southern end of the Bay
of Lima**

Fig. 6 Southern part (*left*) and northern part of the bay (*right*)

2.3.3 Co-seismic EQLs in the Bay of Lima: Video Evidence

The fifth evidence comes from a security video provided by the *"Larcomar"* shopping center, a recent development right at the edge of the cliff overlooking the bay of Lima. The video was one of twelve kept for years by the security officer of the shopping center and generously made available to us for our research. Almost all the cameras were inside buildings and stores but one of them pointed towards the large glass window of the discotheque facing the ocean. Knowing the direction into which the fixed camera was pointing we could deduct by simple ray tracing the origin of a strong and sudden luminescence that was present in two of the video frames. Figure 7 shows the large glass window illuminated by an intense white and light blue flash of light on one of the frames and located as shown on the aerial view

Fig. 7 Luminescence reflected on a glass window pane of a discotheque at *Larcomar* shopping center, Lima and the ray tracing identifying the direction

Glass window pane reflecting the EQL, coming from the red arrow direction, into the camera.

Fig. 8 Security camera #3 at the PUCP campus. Notice the position of camera (*white dot*) and its pointing direction (*black arrow*) towards the west and San Lorenzo Island

of the *Larcomar* shopping center overlooking the ocean. The ray tracing identified the ridge known as "*Morro Solar*" as the direction from where the EQLs arrived.

The sixth and most important graphic evidence comes from security cameras around the campus at PUCP. One of them, camera #3 was fixed, pointing due west at the time of the earthquake. Figure 8 shows the direction (black arrow) of the camera (white dot), looking directly at San Lorenzo Island. This evidence will be further analyzed in the next section.

Fig. 9 Overall scenario of the bay of Lima showing the EQLs studied, observation points and geographical features

Figure 9 shows a compendium of the eyewitness reports and video records of the observations in the bay of Lima. The Google Earth-based view shows the bay of Lima looking south, hence West is to the right in this image, where San Lorenzo and surrounding islands are barely perceptible. This view should be compared with the map shown in Fig. 14, in rather the opposite direction, but the map complements the observed coincidence of the spots where luminescence occurred vs. the geological structures that support the generation of local electric charges.

2.3.4 Analysis of the Observations

The digital video, recovered from storage after almost three years, displays vivid colors, frame numbers, date and timing in milliseconds. It represents a valuable document to the luminous phenomena in the bay of Lima during the Pisco 2007 earthquake. It is also significant in that the direction in which the camera points is easily deductible and coincides with that of the San Lorenzo—El Fronton islands, as seen from the PUCP campus. Four of those frames are shown in Fig. 10 and correspond to the last and most intense EQL.

The intensity of the luminescence was determined by delimiting the area for each of 13 colors, establishing a linear scale and counting square pixels for each frame in the portion of interest with specific software, as depicted in Fig. 11.

A weighted area count gives a sufficiently accurate measure of the intensity of the light versus time. In this way four main time groups were identified. An additional group, group #2, due to light coming from a different direction, was identified. The time, duration, and intensity distribution for each EQL event are summarized in Fig. 12.

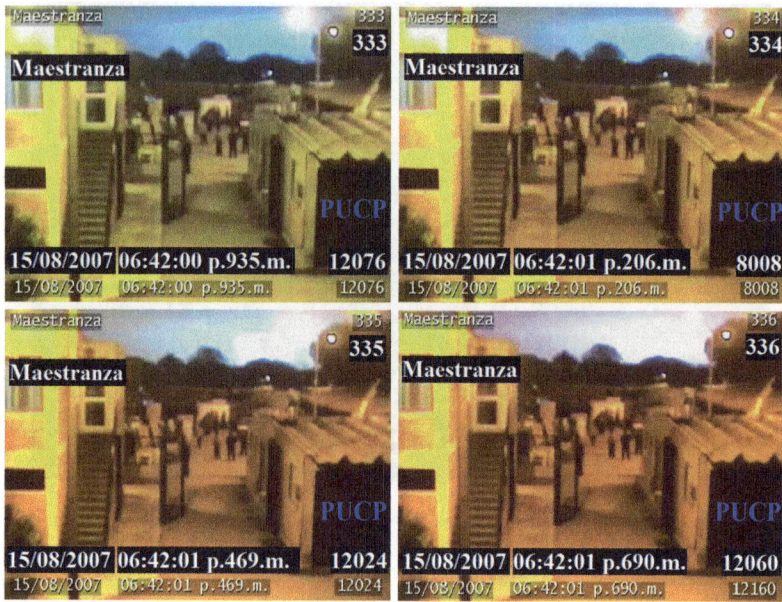

Fig. 10 EQLs in Lima during the 2007 Pisco earthquake as recorded by a stationary security video camera. The frame number is on the *upper right corner* and local time on the *bottom central line*

In order to correlate the times of the EQLs with the earthquake, we took the 3-axis ground acceleration records of the PUCP campus seismometer. Disregarding details of the luminous phenomena, the timing of the EQLs is depicted in Fig. 13, together with the magnitude of the ground acceleration and the time difference correlation of both. We can identify two seismic wave trains corresponding to the two hypocenters that fractured during the Pisco earthquake, approximately 75 seconds apart.

The first wave train started at about 10 sec with the arrival of the first P wave, followed about 20 sec later by the arrival of the first S waves, characterized by significantly higher ground acceleration values. The arrival time difference, about 20 sec, is consistent with the travel time difference between P waves (about 6 km/sec) and S waves (about 3.4 km/sec) and the distance from the earthquake center, 150 km. Excellent correlation is found between the timing of the EQLs and the passage of the S waves leading to peaks in the ground acceleration. The timing of the EQLs with the passage of the S waves is consistent with the observations of voltage pulses recorded by Takeuchi et al. [17].

It is obvious that, for the Pisco earthquake, the EQLs coincide with the arrival of the S waves. This confirms the hypothesis that these luminous phenomena are locally produced, triggered by the extremely rapid, high-amplitude compression and shearing of the rocks during the passage of the S waves.

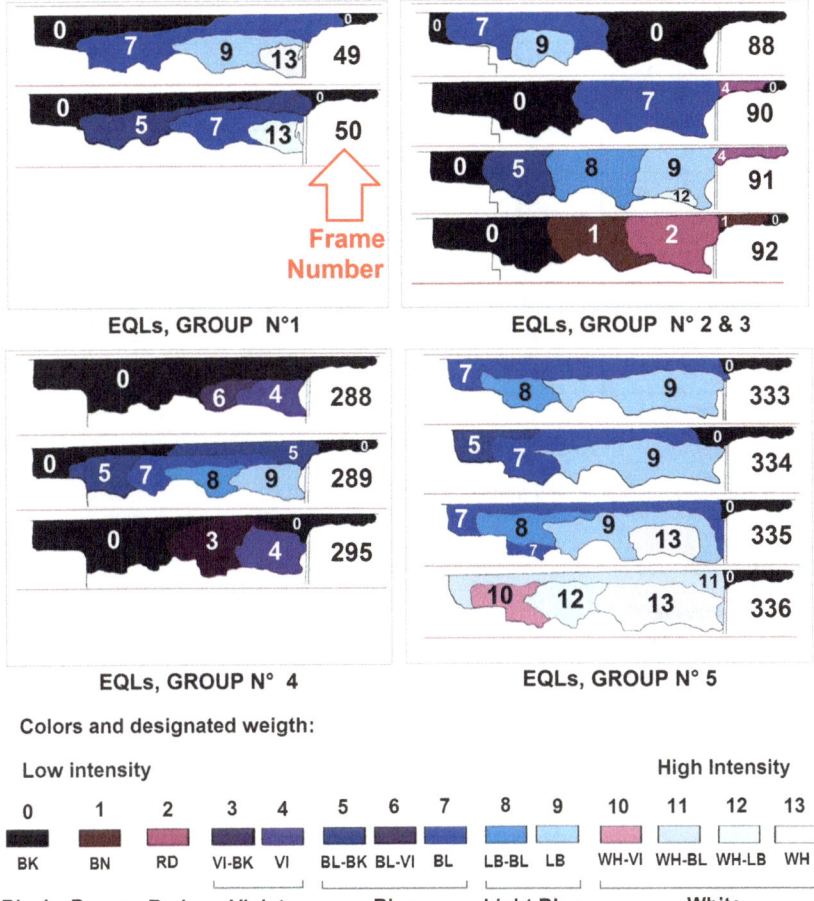

Fig. 11 Luminescence by time frames, colors and areas

2.3.5 Overall View of the Luminous Phenomena

In order to significantly improve the "signal to noise ratio" of the observation, we focused on the phenomena described by witnesses and cameras, across the hills close to the ocean and towards the islands. The observational directions are shown in Fig. 14, with special emphasis on directions, angles of coverage from different observation points from where we have credible reports and video recordings. Additionally, shaded in green, we observe a high coincidence between the areas where the EQLs were video-recorded and reported by qualified witnesses and the geological formation in the bay with mafic igneous rocks of Jurassic and Cretaceous age, a condition that appears to be a prerequisite to the activation of electric charges in the rocks during the sudden compression and shearing during the passage of the S waves.

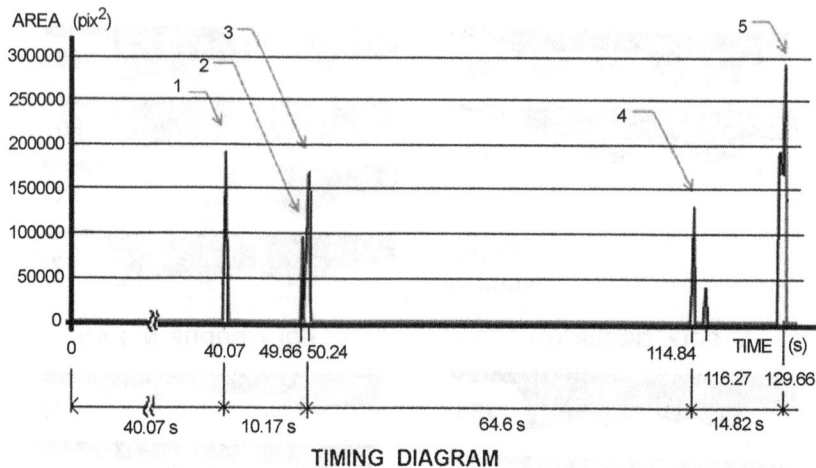

Fig. 12 Resulting morphology and timing diagram for the EQLs

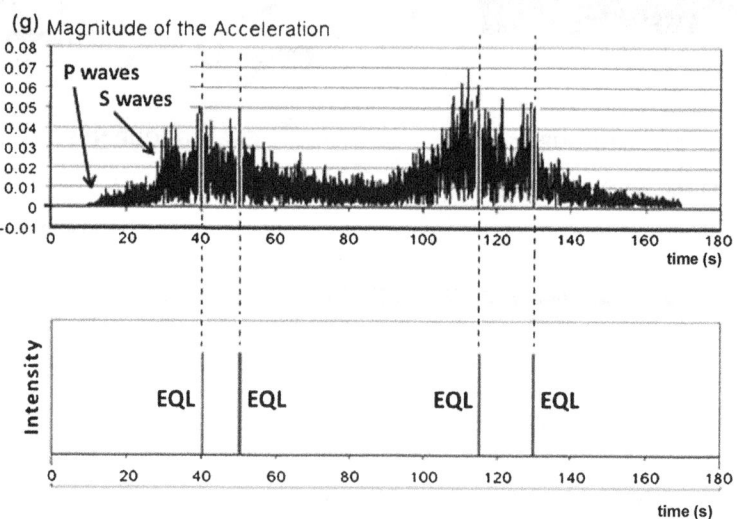

Fig. 13 Time Correlation between the EQLs and the magnitude of the ground acceleration g at the PUCP Campus—August 15, 2007

3 Generation and Transport of Charges

3.1 Generation of Electric Charges Connected to Seismic Activity

Understanding the generation of electric charges in rocks traces back to work done in the late 1980s and early 1990s, when Minoru Freund [18], his father Friedemann Freund and his post-doc François Batllo demonstrated, in a series of elegant labora-

Fig. 14 Map showing the location of the eyewitnesses (airplane pilot, airport control tower observer, officer at naval base on San Lorenzo Island) and of the security video cameras

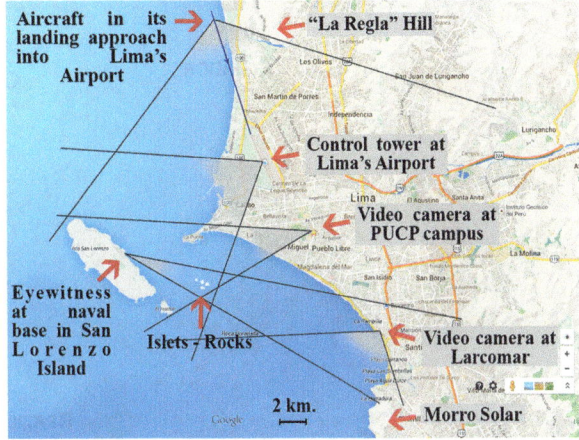

tory experiments, the activation of positively charged electronic charge carriers, first in oxide single crystals, later in silicate minerals. Mino's familiarity with high T_c oxide superconductors, just discovered around this time by Mino's Ph.D. co-advisor Nobel Laureate Alex Müller, helped lay the foundation for the characterization of these remarkable electronic charge carriers, which normally exist in an electrically inactive state. They are now known as positive holes.

The broader geophysical implications of this early work became apparent when, in the early 2000s, Freund [19] and Freund et al. [20] showed for the first time that, when stresses are applied to a slab of rock, granite or gabbro or anorthosite, these same electronic charge carriers become activated as previously identified in single crystal studies. In rocks the positive holes produce a continuous electric current flowing out the stressed rock volume, spreading into and through unstressed rocks, traveling fast and far.

In the previously cited work, Friedemann Freund [1] presents a well-balanced formulation of a solid state theory towards a comprehensive understanding of the generation and transport mechanisms of electric charges in rocks subject to stress. He showed that positive-holes, or p-holes, pre-exist in an electrically inactive, dormant state as peroxy links and that they become activated when stresses are applied. Flowing out of the stressed rock volume the p-holes constitute electric currents. Those currents create electric and magnetic fields, emit ultralow frequency electromagnetic waves, and produce multiple electric phenomena including locally extremely high electric fields capable of ionizing the air, causing dielectric breakdown and bursts of light.

In the earth's crust, the various movements of tectonic plates relative to each other produce great stress levels down to depths of tens of kilometers to hundreds of kilometers. In the Peruvian case, the subduction of the Nazca plate under the continental plate starts at the Peru-Chile trench about 50 km offshore and dips under the continent producing the rise of the Andean ranges along the Pacific coast of the South American subcontinent, from the southern tip of Chile to northern Colombia. Seismic activity is considered "shallow" for hypocenters up to 60 km deep, occur-

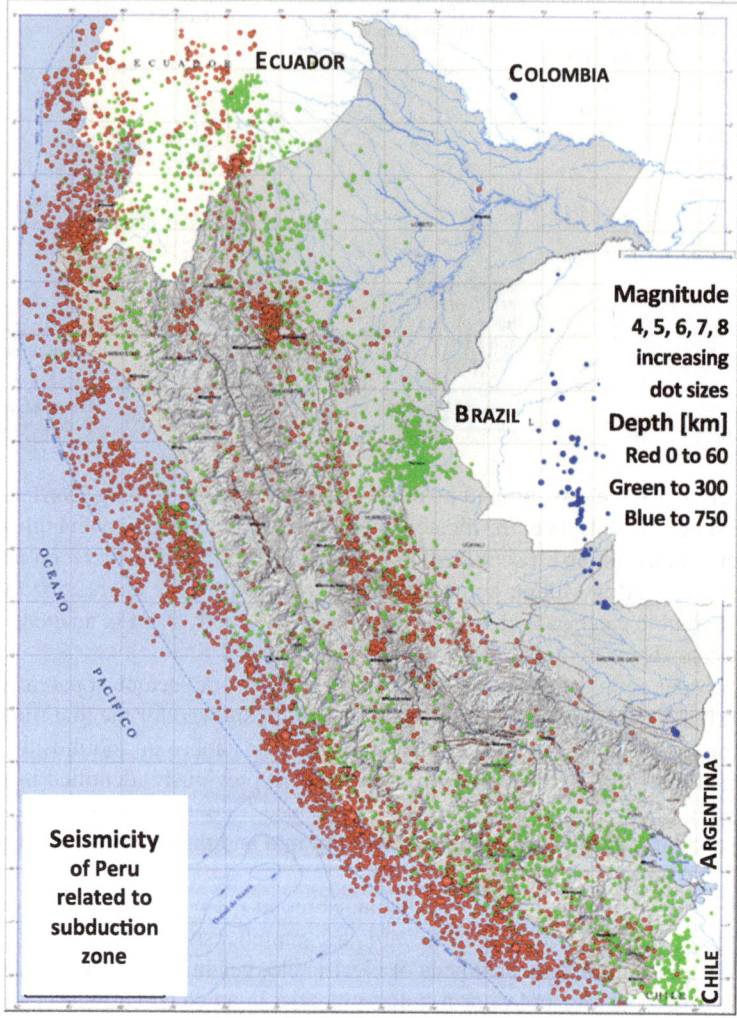

Fig. 15 Seismic activity in Peru plotted by depth and magnitude, 1964–2008

ring mostly along the Peruvian coast, a strip of land that stretches about 100 km from offshore to the rising line of the Andes. Earthquakes occurring between 60 km and 300 km deep in the Peruvian highlands are considered "medium depth" and those from 300 to 750 km occurring in the Peruvian Amazon jungle are considered "deep". Seismic activity from 1964 through 2008 has been plotted by the *Instituto Geofisico del Peru* in these three depths according to their red, green and deep blue color and magnitude according to diameter, as shown in Fig. 15.

Fig. 16 Generation and local
transport of p-holes by
Friedemann Freund

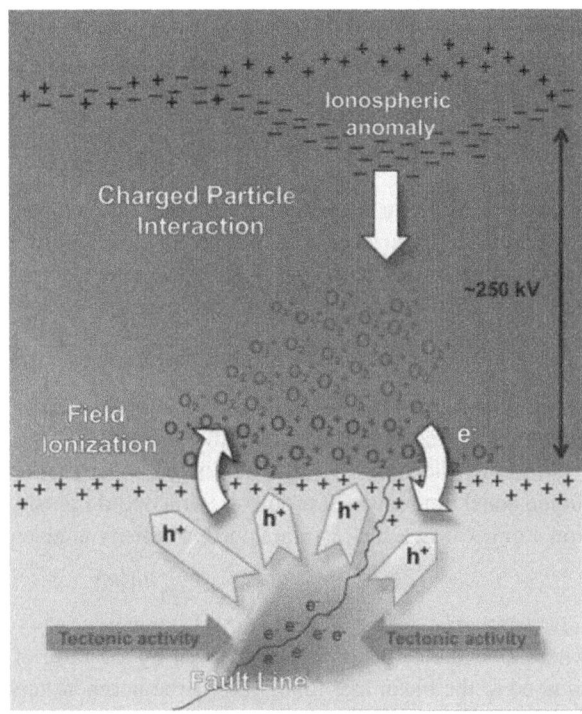

Fig. 16 Generation and local transport of p-holes by Friedemann Freund

3.2 Transport of Electric Charges Connected to Seismic Activity

Freund [1] considers that the generation of p-holes in great quantities produces a migration of these positive charges along the rock, beyond the volume under stress and into the unstressed rock, constituting an electric current of positive carriers subjected to recombination with free electrons. Additionally, this would give rise to emission of infrared radiation and accumulation of surface charges leading to air ionization and to corona discharges due to dielectric breakdown of the air and EQLs. His model is depicted in Fig. 16.

In general, there are several mechanisms by which electric charges will move in a solid:

(a) Drift, i.e. the motion of charges subject to electric and magnetic fields.
(b) Diffusion, caused by the number density gradient of the charges between contiguous volumes of the solid.
(c) Recombination, caused by electrons achieving a lower energy state.
(d) Combined motion of charges in magnetic fields, leading to the Hall effect in semiconductors.

In the case of drift motion of charges, at the presence of an electric field ($\bar{E}$) the lattice should produce a free electron to respond with an acceleration ($\bar{a}$) proportional to the field:

$$\bar{a} = \frac{-q}{m^*}\bar{E}$$

where q is the electron charge and m^* the effective mass. However accelerated motion pertains only to a perfect crystal so in most cases with collisions, chargers move with a drift terminal velocity proportional to the electric field and its constant of proportionality is the electron (or hole), mobility:

$$\bar{v}_n = -\mu_n\bar{E}$$

Motion of charges by drift requires the presence of an electric field and motion is limited by lattice and impurities collisions or scattering.

Motions of charges by diffusion is the consequence of a statistical case of a 3-dimensional random walk problem. For a one-dimensional number density distribution along x, $n(x)$, the resulting current density is given by:

$$\bar{j} = -D\frac{\partial n(x)}{\partial x}\bar{e}_x$$

where D is the diffusion constant given by the ratio of the particle mean free path squared to the mean free time, $n(x)$ is the number density of electrons (or p-holes) and the partial derivative with respect to x denotes a one-dimensional case. In other words, the current density is a vector proportional to the vector gradient of the number density distribution and the negative sign accounts for the fact that charges will diffuse from higher density regions to lower density regions. Hence the presence of an electric field is not needed. Diffusive flow of both negative and positive charges, requires then a density gradient, like a sudden generation of broken peroxy links in great quantities as stated by Freund, to set up high currents of charges rushing to a liberating pointed end. Thus and in a simple way, we can think of diffusion as the transport mechanism for the generation of EQLs.

3.3 A Simple Model for the Peruvian Coast

A simple model that includes the generation, transport and luminescence release is shown on Fig. 17. For EQLs near the epicenter, locally produced charges at the subduction zone where the future hypocenter or rupture point is building up, are responsible for EQLs. As the seismic wave propagates, the local pressure of the S waves leads to the activation of very high concentrations of p-holes, which would then cause the outburst of light from the surface of the Earth due to the dielectric breakdown of the air.

Preliminary analysis of the possibility of generating EQLs with charges that propagate all the way from the epicenter, 160 km away in the case of the Pisco,

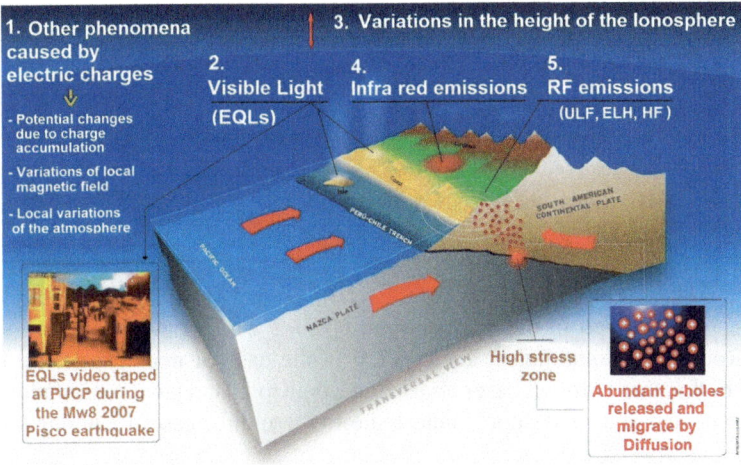

Fig. 17 Subduction plate tectonics along the Peruvian coast and the generation of p-holes giving rise to luminescence at and away from the epicenter stress driven by a seismic wave

Peru earthquake, show that the timing is not compatible with our observations. According to Freund [1] positive holes propagating outward from a source correspond to electrons hopping in the opposite direction from 2-electron (O^{2-}) to 1-electron deficient oxygens (O^-). The maximum speed of propagation is limited by the frequency f of thermal phonons, $\sim 10^{12}$ sec^{-1}, and the inter-atomic O–O distance d, 2.8 Å or 2.8×10^{-10} m:

$$v_{h'} = fd = 10^{12}\,\text{Hz} \times 28 \times 10^{-10}\,\text{m} = 280\,\text{m s}^{-1}$$

This relatively high speed may be applicable for EQLs produced very near the epicenter. However, as the cloud of p-holes spreads further out, it dilutes. Therefore its speed of propagation must slow down. Even considering elevated temperatures deeper in the earth's crust along the propagation paths, providing for a higher conductivity, the speed of p-hole propagation will remain far below the speed of propagation of seismic waves, typically about 4 to 7 km sec^{-1} typically in the Peruvian coast for P-wave and about 3.4 km sec^{-1} for S-waves. Thus, far from epicenters, it is not possible to justify co-seismic EQLs as being produced by p-holes propagating directly from the point of rupture. Therefore, local generation of charges by similar mechanisms must be considered for EQLs coincident with the arrival of the seismic waves. In other words, EQLs must be produced locally, synchronized with the arrival of the seismic waves, specifically S-waves, passing through igneous rock formations close to the surface. Abundant p-holes suddenly formed and creating a strong density gradient, travel by diffusion to pointed formations on the surface, increasing the chances of corona breakdown near the ground and illuminating the surroundings with a fuzzy and rapidly moving flame-like luminescence.

One question that quickly comes up is the reason for the observed time difference between observable electromagnetic precursors and seismic activity, as reported for both EQLs and variations in the local earth's magnetic field in magnetometer

Table 1 Observed EQLs and their relations to earthquakes, for San Lorenzo Is.

EQLs observed			Earthquake			Timing and distance	
Date	Local time	Type	Date	Local time	Mag.	Lead time	Epicenter
1746–1007	02:30	Pre-seismic	1746–1028	22:30	8.6 (est.)	21 days	~2 km NE
2007–0815	18:41	Co-seismic	2007–0815	18:41	8.0	None	~160 km SSE
2012–0727	22:30	Pre-seismic	2012–0729	12:30	4.3	38 hours	~2 km NE

records. As discussed in the next session, up to 2 weeks have been observed in Alum Rock, California and in Tacna, Peru in the case of magnetometers. In the case of EQLs, three distinct cases above San Lorenzo Island have been studied: two pre-seismic and one co-seismic. Table 1 shows these three cases including the lead time.

As can be noticed, pre-seismic EQLs are connected to nearby epicenters, almost precisely in the same place, the oldest one corresponding to the 1746 mega-earthquake (estimated at M8.9 when no seismometers were available, but considered by many seismologists as a M9+ event). We are tempted to conclude that very close to the epicenter, pre-seismic EQLs are observed with short lead-times, of the order of hours, for lower magnitude earthquakes, whereas large magnitude to-be earthquakes show precursor EQLs with lead-times of the order of weeks. This "vicinity condition" to the future epicenter seems to determine the rather proportional factor or lead-time to earthquake magnitude. In the case of the July 29, 2012 earthquake, even though we had the qualified eyewitness report of the EQL observed 38 hours ahead of the earthquake and suspected a possible rupture was in progress, we had no way of determining the time the earthquake would occur. For distant epicenters, tens or hundreds of kilometers away, the effect is still local, produced by p-holes activated in rock volumes that undergo rapid stress build-up without, however, rupturing. The Earth crust is a dynamic system with stresses waxing and waning all the time at different locations, especially in parts of the crust, where seismic activity will eventually occur. Not every local stress build-up leads to catastrophic rupture, hence an earthquake, but episodes of locally high stresses can lead to co-seismic EQLs.

4 Low Frequency Micropulsations in the Local Magnetic Field and Future Research

In 2009, through a cooperative agreement with Quakefinder of Palo Alto, California, deployment of a network of magnetometers, sensitive to ultralow frequency (ULF) electromagnetic (EM) emissions, was started. Dunson et al. [21] have described a similar lead time of about 14 days between ULF pulses and earthquakes, both in Alum Rock, California and in Tacna, Peru. Ten of such induction coil magnetometers have been installed in selected parts of the Peruvian coast, clustered in

Fig. 18 The proposed Peru-Magneto network, showing the existing magnetometer sites (solid colors: *blue*, *orange*, *red*) and the future installations (*black circles*)

order to take advantage of additional triangulation capacity with trustable azimuth observations of sources of ultra low frequency pulsations. One of them has been, just recently, installed at San Lorenzo island, not too far away from the centuries old prison mentioned, with the purpose of studying the propagation of charges and generation of ULF signals. Next to it, automatic recording cameras and other experiments will be deployed shortly. On Fig. 18 the "Peru-Magneto" network is shown, with future stations in black circles and the already installed sites in color-coded solid circles.

5 Concluding Remarks

Luigi Galvani was proven right in interpreting that the muscles of the frog generated electric currents. Over a century later, currents generated by heart muscles were sensed remotely to diagnose heart disease rather than by pressure waves obtained by pressing arteries against bones. Let's be confident that remote sensing of electromagnetic phenomena can provide us with new clues and means to anticipate the *tantrums* of planet Earth.

Acknowledgements   I want to express my recognition to the late Mino Freund, whose dedicated work in various areas of his scientific interest, brought all of us together in this book. From the multidisciplinary symposium held in NASA Ames Research Center on August 10, 2012, it was clear that the broad view Mino had and his enthusiasm for science was the catalyst for a group of scientists in discussing together topics that mixed and matched superbly. My thanks to Friedemann Freund for the honor of being invited to participate.

Part of this work, pertaining to the co-seismic observations in Lima in 2007, was published together with Juan A. Lira. My thanks to Carlos Sotomayor of PUCP for providing me the video recording obtained on the PUCP campus, to CORPAC, the Peruvian Aeronautical Authority and qualified witnesses like air traffic controller Jorge Merino and pilot Giancarlo Crapesi, lieutenant Guillermo Zamorano of the Peruvian Navy, Juan Salas, chief security officer of LAP, the Lima Airport operators and the Larcomar shopping center. I would also like to thank seismologists Daniel Huaco of CERESIS (Regional Seismology Center for South America) and Leonidas Ocola, professor at San Marcos University in Lima. My thanks to Silvia Rosas, Professor and head of the Mining and Geology area at PUCP for her interpretations of the geological maps of the bay of Lima, as well as INRAS research assistants Neils Vilchez and Daniel Menendez for their participation in the experimental work and Victor Centa for the programming and signal processing. Useful discussions were also held with Thomas Bleier of Quakefinder and Friedemann Freund of the SETI Institute, France St-Laurent and Robert Theriault. Pictures of the San Lorenzo prison were taken during a trip to the island supported by the Peruvian Navy. We are indebted to Quakefinder for providing us with magnetometers and to Telefonica del Peru for the donation of the magnetometer now installed at San Lorenzo island, modems and indefinite airtime for getting the data from our magnetometers via cellular 3G data links.

References

1. F. Freund, Toward a unified solid state theory for pre-earthquake signals. Acta Geophys. **58**(5), 719–766 (2010). Institute of Geophysics, Polish Academy of Sciences
2. C.F. Richter, *Elementary Seismology* (W.H. Freeman and Company, San Francisco, 1958), pp. 132–133
3. T. Terada, On luminous phenomena accompanying earthquakes. Proc. Imp. Acad. Jpn. **6**, 401–403 (1930)
4. T. Terada, *On Luminous Phenomena Accompanying Earthquakes*, vol. 9 (Bulletin Earthquake Research Institute, Tokyo University, Tokyo, 1931), pp. 225–254
5. J.S. Derr, Earthquake lights: a review of observations and present theories. Bull. Seismol. Soc. Am. **63**(6), 2177–2187 (1973)
6. G.A. Papadopoulos, Luminous and Fiery phenomena associated with earthquakes in the East Mediterranean, in *Ionospheric Electromagnetic Phenomena Associated with Earthquakes*, ed. by M. Hayakawa (Atmospheric & TERRAPUB, Tokyo, 1999), pp. 559–575
7. F. St. Laurent, The Seguenay, Quebec earthquake lights of Nov. 1988–Jan. 1989. Seismol. Res. Lett. **71**(2), 160–174 (2000)
8. J.A. Heraud, J.A. Lira, Co-seismic Luminescence in Lima, 150 km from the epicenter of the Pisco, Peru earthquake of 15 August 2007. Nat. Hazards Earth Syst. Sci. **11**, 1025–1036 (2011)
9. D. de Esquivel y Navia, *Noticias Cronológicas de la Gran Ciudad del Cuzco, Tomo II*, Fundación Augusto N. Wiese (Banco Wiese, Ltdo., Lima, 1980). Biblioteca Peruana de Cultura Printed in Lima, Peru, Talleres Graficos, P.L., Villanueva, S.A., Chap. XXXVIII, p. 360. The Lima Earthquake, 1746 (translation from Spanish)
10. P.E. Perez-Mallaina, *Retrato de una Ciudad en Crisis. La Sociedad Limeña ante el Movimiento Sismico de 1746* (Pontificia Universidad Catolica del Peru, Instituto Riva Agüero, Lima, 2001), 393 pp.
11. Mercurio Peruano (newspaper), III, 95, folios 239–240, Lima (1791)
12. C.F. Walker, *Shaky Colonialism/The 1746 Earthquake—Tsunami in Lima, Peru and Its Long Aftermath* (Duke University Press, Durham & London, 2008)

13. H. Tavera, I. Bernal, Distribución espacial de áreas de ruptura y lagunas sísmicas en el borde oeste del Perú. Bol. Soc. Geol. Perú **6**, 89–102 (2005)
14. H. Tavera, I. Bernal, H. Salas, El Sismo de Pisco del 15 de Agosto, 2007 (7.9 Mw), Departamento de Ica, Perú (Preliminary Report). Instituto Geofísico del Perú, Lima, Peru, August, 5–47 (2007)
15. L. Ocola, U. Torres, Testimonios y Fenomenología de la Luminiscencia Cosísmica del Terremoto de Pisco del 15 de Agosto 2007. Instituto Geofísico del Perú, http://www.igp.gob.pe/sismologia/libro/trabajo_19.pdf (2007)
16. Institute for Radio Astronomy, Pontificia Universidad Catolica del Peru (INRAS-PUCP) Video of Earthquake Lights recorded at the PUCP campus, Lima, Peru, during the Pisco earthquake of August 15, 2007. http://www.pucp.edu.pe/inras/peru-magneto/mod-3cam-3.exe
17. A. Takeuchi, Y. Futada, K. Okubo, N. Takeuchi, Positive electrification on the floor of an underground mine gallery at the arrival of seismic waves and similar electrification on the surface of partially stressed rocks in laboratory. Terra Nova **22**(3), 203–207 (2004)
18. M.M. Freund, F. Freund, F. Batllo, Highly mobile oxygen holes in magnesium oxide. Phys. Rev. Lett. **63**, 2096–2099 (1989)
19. F. Freund, Charge generation and propagation in rocks. J. Geodyn. **33**(4–5), 545–572 (2002)
20. F.T. Freund, A. Takeuchi, B.W. Lau, Electric currents streaming out of stressed igneous rocks—a step towards understanding pre-earthquake low frequency EM emissions. Phys. Chem. Earth **31**(4–9), 389–396 (2006)
21. J.C. Dunson, T.E. Bleier, S. Roth, J.A. Heraud, C.H. Alvarez, J.A. Lira, The Pulse Azimuth effect as seen in induction coil magnetometers located in California and Peru 2007–2010, and its possible association with earthquakes. Nat. Hazards Earth Syst. Sci. **11**, 2085–2105 (2011)

Rock Softening with Consequences for Earthquake Science

Claes Hedberg

1 Introduction

In recent years, my group at the Blekinge Institute of Technology in Sweden has used acoustics extensively to study materials under stress. We observed some unexpected phenomena. Because acoustics afford an extremely sensitive tool, we have been able to probe these effects in detail. We found some of them to be almost universal, occurring in all materials we have tested to date such as organic composites, concrete, rocks, even metals. Our results carry broad implications for a number of fields. In particular, we became aware that our results may prove invaluable for elucidating certain properties of rocks, which have previously been poorly understood but which may be important for understanding pre-earthquake processes. As part of our effort to identify other researchers, who might be interested in this field, we heard of Minoru Freund and his father Friedemann Freund, who were working at NASA in California on fundamental solid state processes underlying pre-earthquake phenomena. With a visit to the Golden State in 2010 began an intense contact, highlighted by many insightful discussions and soon followed by a NASA-funded project on the softening of rocks. Mino—as everybody called him—was its Principal Investigator. Unfortunately, Mino's broad and diverse scientific career was cut short barely 12 months later by an incurable brain tumor and the project, which we had started, became the last he was able to carry through.

In this chapter, I will discuss some of the unexpected and initially seemingly inexplicable observations that we made at Blekinge and the hypotheses, which evolved. In deference to Mino's contributions my focus will be on the possibility of deriving information on the approach of a major earthquake.

C. Hedberg (✉)
Blekinge Institute of Technology, Karlskrona, Sweden
e-mail: claes.hedberg@bth.se

C. Hedberg
SETI Institute, Mountain View, CA, USA

F. Freund, S. Langhoff (eds.), *Universe of Scales: From Nanotechnology to Cosmology*,
Springer Proceedings in Physics 150, DOI 10.1007/978-3-319-02207-9_17,
© Springer International Publishing Switzerland 2014

In the following section, I will summarize the results of our nonlinear acoustics studies. Thereafter, these findings will be used to form hypotheses related to the earthquake process. A short description of the possible consequences will follow, including a proposed mechanism for earthquake triggering. Lastly, I will consider results from recent experiments and close with a few open questions requiring further research. Before embarking on these advanced topics, however, we must first define some general principles of prediction as well as some common understandings crucial to materials science.

In principle, predictions may be based either on certainty or on probabilistic grounds. Predictions based on certainty require that the system under consideration passes a point-of-no-return, that one has knowledge of the process in question, that one can describe it, and that one has the ability to detect the process. In practice, this means that, in order to arrive at prediction based on certainty, the system must send out a specific signal or some measurements must be possible that produce specific parameters. With regards to earthquakes, predictions based on certainty are impossible, but predictions based on probabilities are being widely used. Within seismology the probability of a future earthquake along a given section of a tectonic fault is constructed from experiences gathered by analyzing past earthquakes. Some researchers outside seismology have suggested that, prior to major earthquakes and also volcanic eruptions the Earth produces electromagnetic signals, which have predictive power [1–8]. Not everybody agrees, however, that this is a valid approach and many call into question whether non-seismic precursors even exist and can be shown to be produced in all cases (i.e. [9, 10]).

Our work reported here is based on three basic facts common to all of materials science.

(i) The state of any material can be changed by a number of external influences. Mechanical impacts, ultrasound, heat, magnetic fields, electricity and humidity, all can alter a material's macroscopic properties reversibly [11–15].
(ii) Various external influences may act simultaneously and thereby explain some effects, which cannot be described by any one influence. Thus, we must consider that external influences can be additive with respect to type of influence as well as time or duration.
(iii) We must admit that, when one material property changes, other properties inevitably also change. For instance, a mechanical influence changes both the elastic and electric states, as does an electromagnetic influence.

Finally, a difficulty familiar to all experimental sciences is the need of be able to distinguish between causes and effects. One has to always ask: Is the observed signal due to the process itself? Is it a trigger? Is it both? In addition we may ask ourselves whether a given signal can be regarded as a precursor?

2 What We Know from Acoustic Measurements on Rocks

For several years, linear and nonlinear acoustics have been used for sensitive measurements of the material state. Fundamental research on rocks and other solids has

Fig. 1 The configuration that
has been used for the material
monitoring by acoustics. The
material samples were inside
the climate chamber (see
Fig. 2)

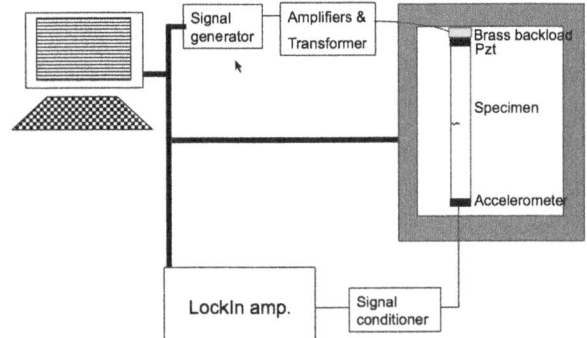

Fig. 2 *Left*: The climate chamber. *Right*: A few different rocks and a couple of other test samples

been done in order to gain insight into the fast and slow dynamics of a material's
reaction to external influences. Acoustic resonance measurements have, in principle, a very high accuracy in the determination of the resonance frequency f_r, and
therefore of the sound velocity c of the objects (as $c = L/f_r$, where L is the length
of the object). A change in sound velocity c is connected to a change in the elastic
modulus E, through $c = E/\rho$. Because changes in temperature and humidity also
affect rocks, many of the experiments were done in a climate chamber such as the
one shown in Fig. 2 (Left) where temperature and humidity can be controlled.

The rock sample has usually been a cylinder with a diameter of 30 mm and a
length of about 300 mm. Most often, the external influence (the conditioning) was
made with an acoustic wave so that both, conditioning and probing, was based on
acoustics. The acoustic transmitter was a power PZT glued to one end of the cylinder, while a smaller sensing PZT was glued to the other end as depicted in Fig. 1.
Resonance frequency tests were made by changing the input frequency in a step-

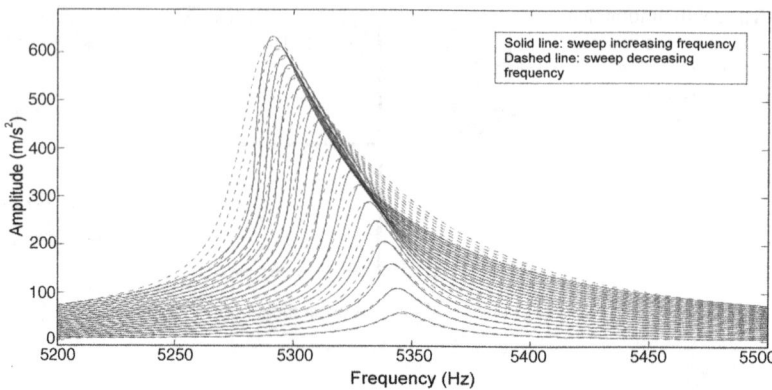

Fig. 3 A number of normal frequency sweep done upwards (*solid lines*), and downwards (*dashed lines*) for several different acoustic input amplitudes

wise fashion, monitoring the output amplitude of the wave with a lock-in amplifier [16–18].

Figure 2 (Right) shows some of the samples that we have tested. Several were rocks plus other materials. Changes in the resonance frequency of a mode implies a change in elastic modulus. One characteristic of the behavior is that the elastic modulus always decreases for any external disturbance to the state [19, 20].

Figure 3 shows the measured wave amplitude for a number of different input amplitudes. There are two curves for each amplitude. The first, done first, is where the frequency is increasing (the solid line). The second one, done directly afterwards, is where the frequency is decreasing (the dashed line).

One can recognize two major effects. First, the resonance frequency decreases with amplitude, and the resonance curves change their shapes. Second, the downward sweep curves differ from the upward sweep curves. The conclusions to be drawn from this observation are: (i) the material is nonlinear; (ii) its elastic modulus decreases with the acoustic amplitude; and (iii) there is a time dependent hysteresis.

This time hysteresis is sometimes called Slow Dynamics, because recovery after disturbances can be measured over minutes, hours, and even days [20, 21]. One might at first suspect that it is just a temperature effect. But it is not, as can be seen, for example, from the data in Fig. 4 where the resonance frequency shift from the acoustic influence at strain 2 μ is equivalent to a temperature change of 5 K. But this was not the case when the temperature was controlled [22].

Another example of the time-dependent hysteresis effect was obtained for the configuration depicted in Fig. 5. The sample monitored here is not a rock but a thin strip of LDPE-Paper-Al. Here, the external influence is the change in strain (length of sample), as seen in Fig. 6. The stress was measured by the tensile test machine and the resonance frequency was measured with a doppler laser vibrometer (C in Fig. 5) [23].

When the strain 'jumps' up to the high values, the stress increases and then starts to decrease slowly. This means that the elastic modulus also slowly decreases. For

Fig. 4 Tests done in the climate chamber for three different temperatures. It is seen how an acoustic strain amplitude of 2 μ in this case is equivalent to a temperature change of 5 °C

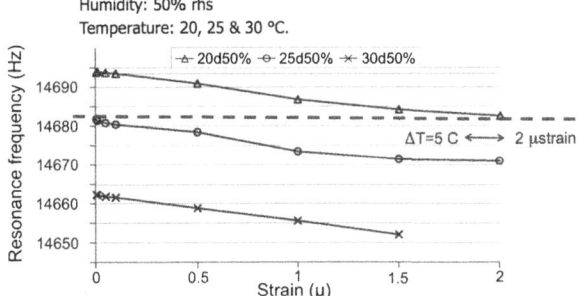

Fig. 5 Non-contact configuration of thin strips (*D*) in the tensile test machine (*H*)

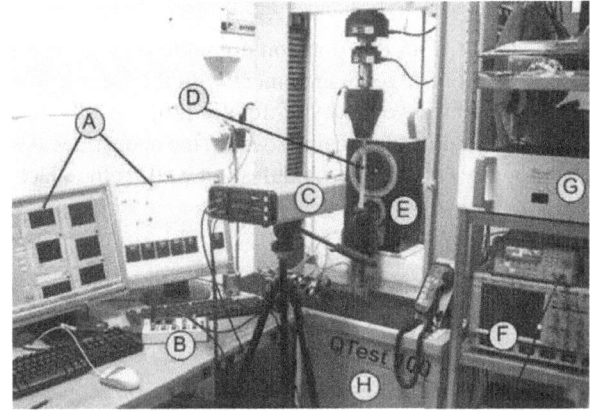

Fig. 6 The strain conditioning (*lowest curve*); and the measured resonance frequency, and the stress for a thin strip of LDPE-Paper-Aluminum

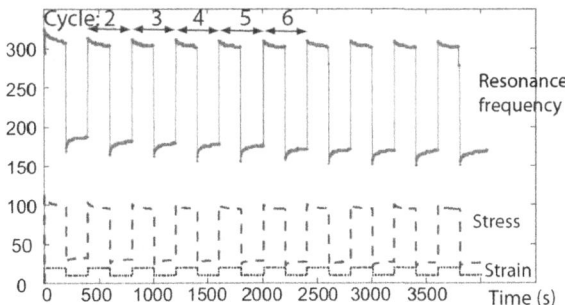

this configuration, the sound velocity is $c = \sqrt{T/(\rho h)} = \sqrt{E\varepsilon/\rho}$, where c is the velocity of the bending wave, T is the tensile force per unit length, ρ is the density, h is the thickness, E is the elastic modulus, and ε is the strain [24]. The strain is just a step function switching between two values. Similar results were obtained on rocks by TenCate et al. [25] using ultrasound as the conditioner.

If one material state parameter changes, other ones should follow. In order to show this, one can make a measurement of the electric capacitance as a function of acoustic amplitude.

Fig. 7 A change in the acoustic amplitude in a granite rock results in a change in the electric capacitance. The measurement (*solid line*) and the theory (*dashed line*) agree well

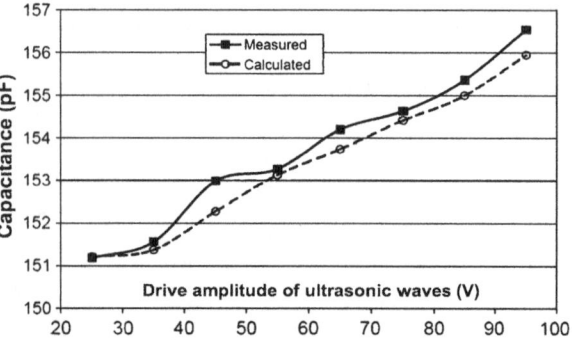

The result from this done on a granite rock is seen in Fig. 7, where the experimental result is in good agreement with the theory, which was developed based on a generalized Debye theory of dipoles [26].

An external mechanical influence (the acoustic wave) changes both the mechanical property (the elastic modulus) as well as the electric property (capacitance or dielectric constant).

2.1 What We Know So Far

Having completed these tests, the science of acoustics allows us to deduce many additional facts about the materials studied, some of them expected and others quite novel and significant. We know with certainty that mechanical influences like ultrasound, impacts, or stress-strain motions decrease the elastic modulus. Likewise, other influences, like electromagnetic waves, electric currents, and of course heat also decrease the elastic modulus. The decrease in one material property such as the elastic modulus is accompanied by changes in the other properties such as permittivity. The changes we measured depend on the level of a particular type of damage in the material. Significantly, however, these changes are fully reversible. It is intriguing to note that these changes persist for some time after the external influence has ceased. Moreover, the recovery rate may be further divided into a fast part and a slow part, where the slow dynamic recovery may take minutes, hours, or even longer. In subsequent testing, additional influences add to the earlier effects, at least in part. We have shown that this behavior is general for all kinds of hard or soft materials, ranging from rocks to metals to organic polymers and composites, and that a relatively small amount of energy causes a large change in material properties.

3 Hypothesis

We have seen from what was described in the previous section that many materials (actually all that we have tested) have the same type of behavior: The material state changes when externally influenced, and there after the material slowly recovers. It has been seen that materials with higher damage levels (e.g. materials with more cracks) show a larger response. The earth consists to a large degree of rocks, which inherently have a high damage level. One can expect similar behavior of the materials in the ground when subjected to external influences. Thus, we hypothesize that the elastic modulus and other state parameters of subterranean rock would be influenced in a similar way as the rocks are in the laboratory experiments.

4 Possible Consequences

Our hypothesis is not without serious consequences for earthquake science. We have established that a small amount of energy yields large changes in material properties. Thus, it follows that triggering an earthquake does not require large energies. Any decrease in elastic modulus must imply a decrease in wave velocities. We also know that the effect from influences of different types is partly additive, as is the effect from influences in time, up to on the order of hours at least. Likewise, recovery takes a certain time, and what happens during that time opens a possibility that specific signals or material parameter changes may be detected. Recovery is reversible. No increase in damage (for example, an increase in micro-cracking at or near the prospective hypocenter) is required. Another consequence is perhaps less obvious. At least in principle, no increase in stress is required. This consequence will become apparent in the suggested simple scheme to follow.

From what has been seen in all the materials we have tested, we can assume that the more damaged the rocks are, the more pronounced the effect. The earth is a dynamic system, constantly subjected to stresses and strains, exposed to influences external to the ground, influences from normal processes in the ground, and influences from triggered processes in the ground. Among these influences, which might contribute in affecting the material of the Earth's crust and upper mantle, we may include: electromagnetic waves ranging from the effects of sunlight to the effects of solar storms; mechanical waves such as seismicity and ocean waves; induced telluric currents and other electrical effects; solar storms; stress-strain changes due to all forms of tides, tectonic plate motions or elasticity decrease; and, of course, heat.

4.1 A Scheme of Recovery for a Simple Model

In order to illustrate the viability of the claimed consequences, a simplified model of an earthquake fault will be used in Fig. 8. We assume two regions of ground that have been moving relative to each other for a few years so that stresses have built up in the Fault zone (in black). In Fig. 9 we divide the stresses into components parallel to the fault plane (shear stress) and components perpendicular to the fault plane (compressive stress). With some basic assumptions we consider that the compressive forces press the fault together, while the shear forces try to make the fault

Fig. 8 Two tectonic plate sections moving relative to each other

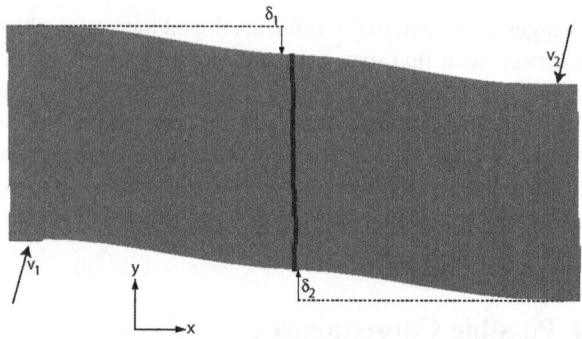

Fig. 9 Schematic representation of the fault zone and the outer regions. Stresses: transverse and parallel to fault. And Shear versus Compression effects. The *thick vertical arrows* indicate the fault shear forces, and the *horizontal* ones indicate the fault compressive forces

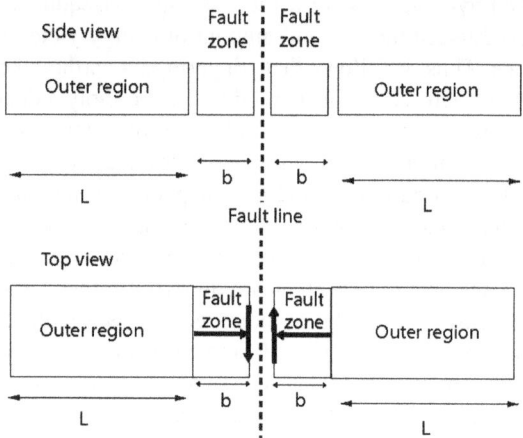

slip. The width of the fault zone b is taken to be on the order of ~10 meter, while the width L of the outer region is on the order of ~10s of kilometers. The boundaries of the Outer regions are fixed in the current time perspective. The Fault zone is assumed to have a larger degree of damage (more cracks) than the Outer region.

A few other assumptions are:

I. The rupture takes place at the fault gouge.
II. It is the fault zone that critically restrains the rupture.
III. The fault zone is more damaged than the rest, i.e. it is more affected by external influences and has a longer recovery time.
IV. The stress level in the fault zone is predominantly determined by (a) the strain acting on it from the surrounding (outer) region; and (b) the contraction or expansion of the fault region; and (c) the change in the elastic modulus of the fault region.

Figure 10 describes the stress-strain diagrams of the material states in the Outer region on the left, and the Fault region on the right. The states are shown for three stages.

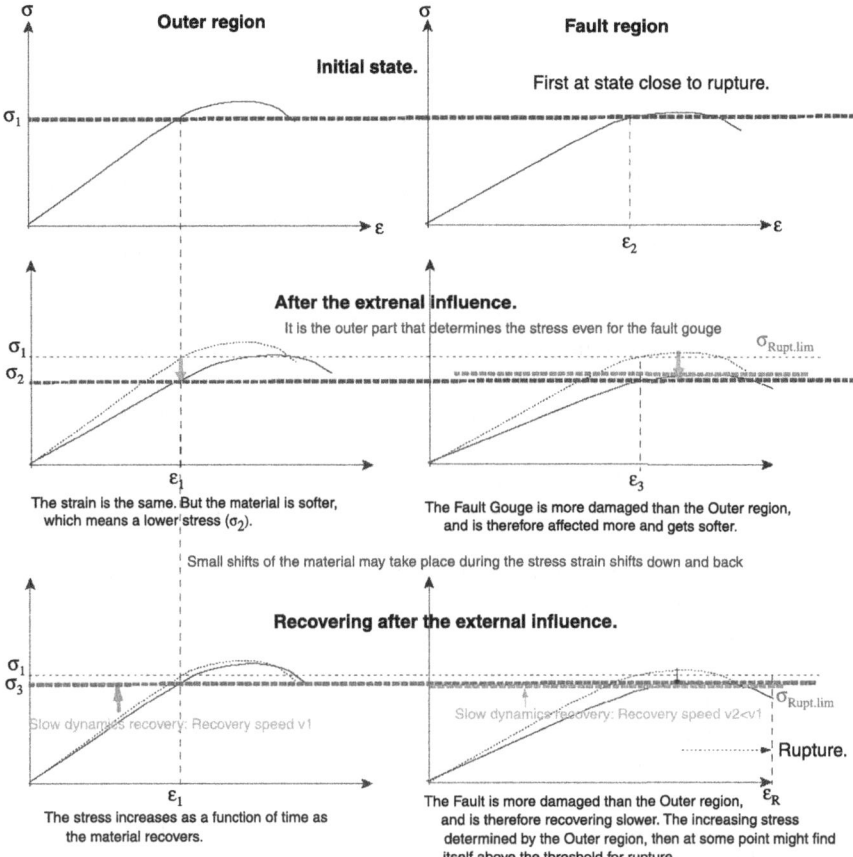

Fig. 10 The combined external influences during the recovery phase moves the strain over the fault zone all the way above the current rupture limit

The top row shows the starting states, when the Fault zone is already relatively close to rupture. The elastic moduli of both regions are given together with their stress-strain curves. The overall stress for both regions is given by the stress of the Outer region, because it is much larger but there is still no major motion.

In the first step of the process, the elastic modulus decreases from E_{01} to E_{02}. This means that the stress-strain curves of the materials change so that they are lower than before. A basic estimation for the new stress level can be made using the formula $\sigma = E \cdot \varepsilon$, where ε is the strain. No major motions means that the strain of the Outer region $\varepsilon_1 \approx$ constant $= \sigma_1/\varepsilon_1 = \sigma_2/\varepsilon_2$ and that the stress σ_2 must also decrease (because $\varepsilon_2 < \varepsilon_1$). Thus, we have a situation where the stress σ_2, which affects both regions, is lower and that the stress-strain curves for both regions are also lower. This situation is shown in the middle row. There is still no rupture.

As the recovery process begins, the Outer region recovers faster than the Fault region because the recovery speed is dependent on the amount of damage. And it is (still) the Outer region which determines the stress in both regions. Therefore the stress will recover faster than the stress-strain curve of the Fault. And when the stress passes the rupture threshold, the rupture takes place.

The effect of triggering is not immediate. Instead there is delayed triggering. The softening of the rocks as a mechanism for triggering is one of two rules spelled out by Hough [27]: "To generate an earthquake, one of two things must happen: (1) stress must build, either gradually or suddenly, to the point that faults reach a breaking point; or (2) the fault must somehow weaken such that rupture is facilitated."

A stress decrease before an earthquake is contradictory to Rule 1. Indeed, many authors believe that stress must increase until the release takes place at the moment of an earthquake, and that both stress and the number of micro-cracks must increase just before an earthquake. However, in many documented cases, measurements of the primary and secondary wave velocities have provided evidence that the stress decreases before earthquakes (Gao [28]). This is what our findings also support for the free boundary rocks: the stress decreases, and this can be part of what triggers earthquakes.

Figure 11 is very similar to Fig. 10. It describes the effects of baseline disturbances that take place at a diurnal rhythm. Examples of these recurring influences are gravitational tidal effects, heat and other electromagnetic radiation from the sun, and telluric currents on a background of teleseismic waves. The baseline disturbances affect the rocks by repeatedly covering strain excursons $\Delta \varepsilon$. Because this is present in principle every day, it can be regarded as a determinant for the noise baseline level. In order for a specific effect to be acting as a triggering, it should be above this threshold level.

Again, when an external perturbation occurs, the elastic moduli of both regions decrease, close to the fault and far outside. At the same time, assuming the boundaries of the regions do not change, the strain field remains approximately the same, which must lead to a decrease in stress. This means that we have competing actions from (1) the stress and (2) the material parameter curve of the Fault region. Because the Fault region has more damage, it is affected to a higher degree by the external influences. The stress might then grow beyond the critical value of the Fault region. The external influence may consist of a combination of influences such as seismic and electromagnetic, which when added to a ripe state of the plates, can trigger the event.

5 Support for the Hypotheses and Consequences

Now, we may consider a few examples of measurements, which support the processes described above. Some tests show how different influences in the lab yield a softening of a small rock beam. A recovery process then follows. But we will begin by considering a large composite beam made of glass fiber reinforced plastics and steel.

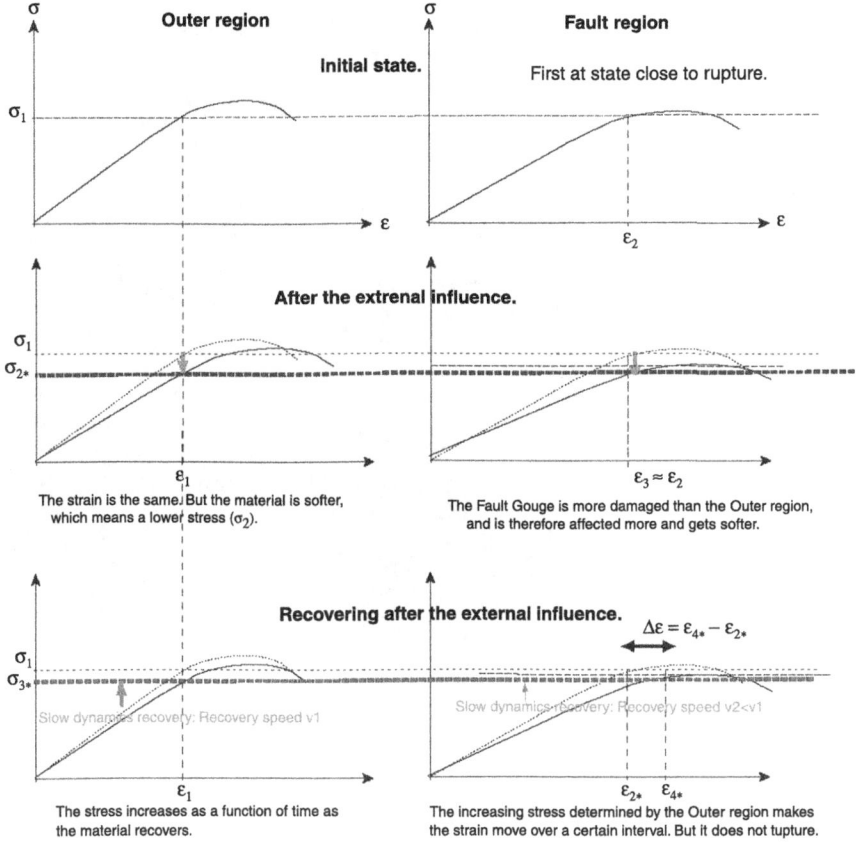

Fig. 11 The daily external influences move the strain over the fault zone across a certain interval. This is why triggering external influences usually must be above a certain threshold

5.1 Test on a Large Composite Beam in March 2011

While performing a 20-day long fatigue test on a 4-meter long composite-steel beam on March 3–22, 2011 (Fig. 12) we noticed a sudden large decrease in the elastic modulus as shown in Fig. 13 [29]. The influence seemed external. What could have been the reason for the sudden softening?

The beam was monitored in two ways. The elastic modulus was measured through the load force and the displacement. Damage was monitored by a nonlinear acoustic technique.

The two downward peaks in Fig. 13 indicated a sudden increase in the displacement, apparently due to softening. After this anomaly, the beam recovered and returned to a stiffer state without any action from our side. Within a few days, slow recovery brought the curve back to its previous linear slope.

Fig. 12 Test of composite
beam in 2011

We were aware of the possibility of a material becoming suddenly softer, because this type of fast decrease and slow recovery had been seen during many other tests where the causes were mechanical, from either ultrasound or applied strain [16–19], but such an effect was not expected in this situation.

Here the beam was subjected once every second to a dynamic load force of more than 100 kN, and it was highly improbable that any other external mechanical force could surpass this influence. The temperature and humidity were stable. The cause for the observed anomaly could not be mechanical and had to be external.

The first dip is at 674,000 cycles, which occurred at around 10 UTC on March 10, 2011. The second dip (which is the lowest) is at about 710,000 cycles, which occurs at 20 UTC on March 10. The damage level monitoring (Fig. 14) was showing high fluctuations in values some time before and after the sudden decrease in stiffness, and for a brief period the acquisition system shut down completely. Neither the sudden softening of the beam, nor the shutdown of the acquisitions system could have been caused by mechanical means, but indicated possible external electrical or electromagnetic disturbances.

In contrast to the elastic stiffness, the average damage level did not decrease; it only showed a higher variance. As with the softening of the beam and the shutdown of the system, the most probable cause seemed to be electric or electromagnetic influences, and the first thing that came to mind was an influence from the sun, as solar wind effects on electronic instruments and power lines are well documented.

Fig. 13 The stiffness of the composite beam as measured by the loading force and the beam deflection

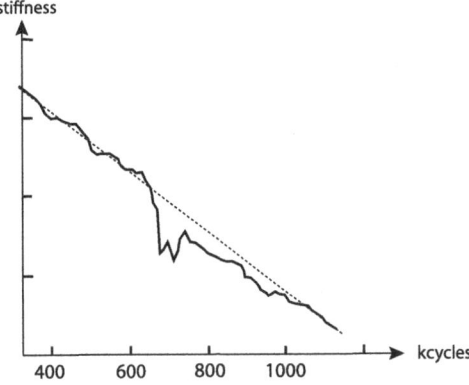

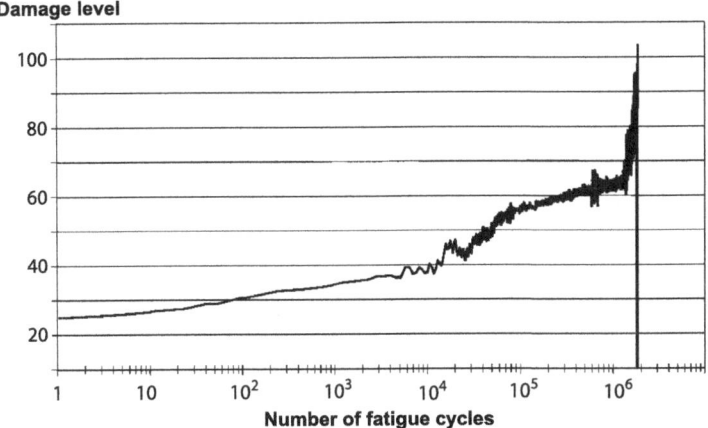

Fig. 14 The damage level of the composite beam as monitored by acoustic techniques during the fatigue test

As it happened, the sun really had had major eruptions on March 7 and 9[1] and the solar wind reached the earth later on. Satellite-stationed charged particle sensors recorded a maximum at around 6 UTC on March 10, 2011 [30]. The ACE satellite gives about one hour advance warning of geomagnetic storms http://www.srl.caltech.edu/ACE/ace_mission.html.

We hypothesize that it was a solar wind influence that was detected through our 20-day mechanical fatigue test, because the observation fits quite well in time with the recorded geomagnetic measurements, and because we already know that electric properties correlate to the elastic stiffness. This was later tested because of the expected connection between the elastic state and the electrical properties [26, 31, 32]. As we were not prepared to look for it, we do not know the exact way that the beam

---

[1]http://science.nasa.gov/science-news/science-at-nasa/2011/14apr_thewatchedpot/;
April 19, 2011.

was influenced. It could for example have been by electromagnetic fields created by ions, or it could have been an electric current coming in through the ground. Because the material weakening coincided with the largest solar eruptions in years, the thought occurred that it might also have had an effect on the magnitude 9 earthquake in northern Japan 20 hours later on March 11, 2011 at 5.47 UTC, due to the Earths crust being weakened. In fact, statistical correlations between earthquakes and solar activity have been reported and discussed many times (e.g. [33]). Mechanisms that have been proposed include (i) the Lorentz force, which results from telluric currents induced in the solid Earth by currents in the ionosphere overhead and leads to a mechanical torque; (ii) magnetostriction of the rocks giving rise to a horizontal force; and (iii) the Einstein-de-Hass force also resulting in a torque similar to that of the Lorentz force [34].

We propose that mechanisms exist which lead to a decrease in the elastic properties of rocks in the Earth crust and affect their weakening and subsequent recovery. However, it must be stated categorically that solar activity alone cannot be used as a predictor of major earthquakes. There are major earthquakes without solar flares, and there are large solar flares without major earthquakes.

5.2 Cantilevered Beam Experiment

In order to show that a beam can indeed be softened and recovery we set up a new type of laboratory experiment, where a rock beam was clamped horizontally at one end, as shown in Figs. 15 and 16 with the deflection of the free end being monitored by a laser-based displacement sensor.

The beam was activated with an ultrasound wave transmitted from one end of the beam. The beam started to deflect instantly as the ultrasound was turned on but then kept sagging more slowly for more than 20 minutes as illustrated in Fig. 17. When the ultrasound was turned off, the beam instantly rose again, continuing to rise more slowly until it eventually returned to approximately the same level as before [35].

To demonstrate that the electrical properties of the beam changed at the same time, we recorded the current as shown in Fig. 18. There is a clear difference between the electric and the elastic properties with and without the beam being exposed to ultrasound. The electric property (upper plot) shows no time-dependent hysteresis: it simply increased and returned immediately to its former level as soon as the ultrasound was turned off. The elastic property, i.e. the deflection, shows a slow return to the initial value.

Fig. 15 The cantilever beam configuration. It is simply a beam held in one end, and a laser measures the deflection of the other end

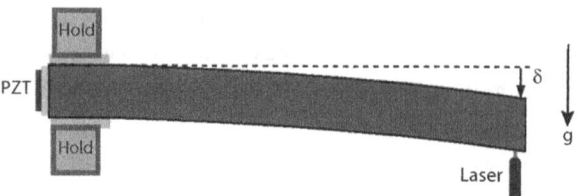

Fig. 16 The first experimental lab set-up of the cantilever beam, according to the scheme in Fig. 15

Fig. 17 The deflection of the beam free end (*solid*), and the ultrasound amplitude (*dotted*). The deflection kept increasing during the time the ultrasound was on, and it started to recover as soon as the ultrasound stopped

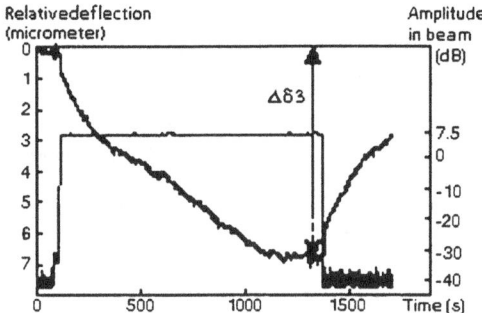

5.3 Thresholds and Delay Triggering

The process in Fig. 11 indicates a strain baseline level, supported by both laboratory and field observations. For tests inside the climate chamber, the threshold was lower [16, 36]. Slightly depending upon the test, the threshold ranges between an acoustic strain of 0.05–0.12 µ as exemplified in Fig. 19.

From field data, the necessary strain level of triggering seismic waves seems to lie in the region of 0.1–0.25 µ [37]. In this connection Pollitz [38] writes that: "The above considerations provide only partial answers to the question of why the 2012 event triggered so many remote large aftershocks". The dynamic strain threshold level was here taken as 0.1 µ strain for at least 100 seconds. The remaining contribution to this increase may (according to our assumptions of this chapter) may relate to the particularly strong solar wind [30] with peaks at 18 UTC on April 10, 2011 and at 10 UTC on April 11, 2011. The solar wind effect might have been contributing to both the original triggering of the M9 earthquake of March 11, 2011 as well as of the after shocks.

The total effect baseline is presumed to be higher in the real earth than in the lab, because the lab tests are usually done at constant T, no electric currents, no extra vibrations, no direct sunlight etc. The existence of the threshold there presumably indicates the level of constant background influence of the temperature itself. Re-

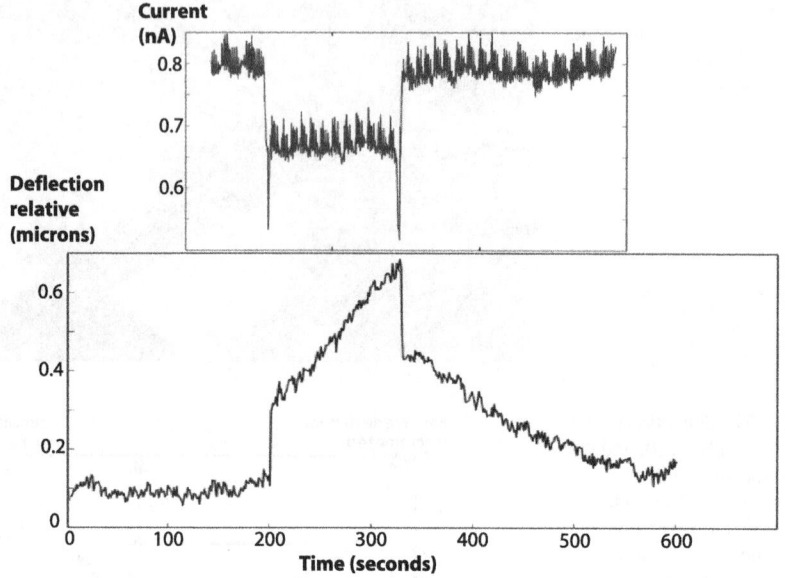

Fig. 18 The electric current (*top*); and the deflection (*below*). The ultrasound was on during the time that the current was at the lower value

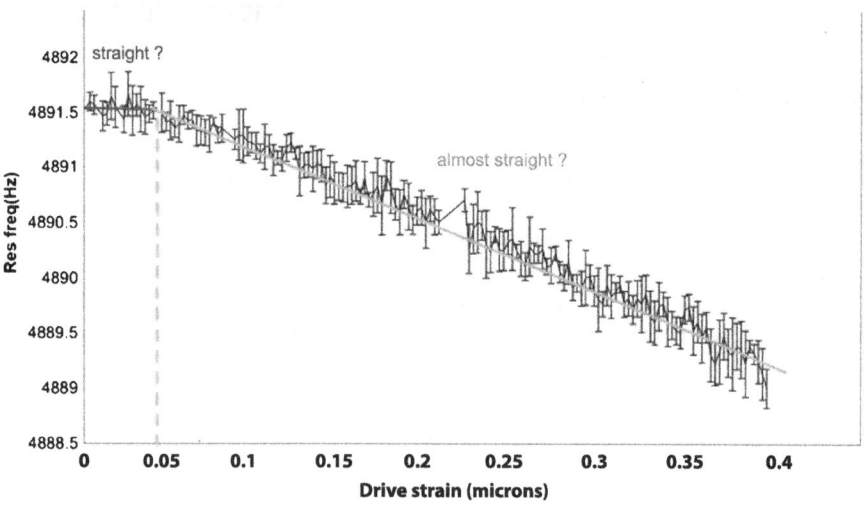

Fig. 19 Resonance frequency shifts at an ultrasound strain of 0.05 μ

maining parts of triggering could be due to other causes such as telluric currents, cf. [39–41].

Delayed EQs caused by seismic triggering tend to occur within 24 hours [37]. From comparisons of earthquake occurrences and solar wind data, it was concluded

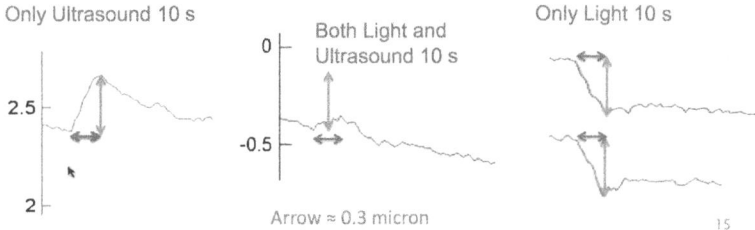

Fig. 20 The deflection of a rock beam for three conditionings. *Left*: ultrasound; *middle*: ultrasound and light; *right*: light

that, whenever there is a temporal correlation between major earthquakes and large solar storms, the quakes never seem to occur before or at the maximum of the geomagnetic anomaly, but afterwards within 24 hours, i.e. with about the same time lag as for seismic triggering. Both of these phenomena could be due to a process, which includes a reversible recovery of the mechanical properties of the rocks.

Aftershock triggering by mechanical causes has been proposed and studied by several authors beginning with Benioff in 1951 [42]. A more recent proposal of successive (shear) modulus weakening by seismic waves, based on the slow dynamics of materials, was made by Johnson and Jia in 2005 [44], whose scheme indicates additive softenings. Triggered creep as a mechanism has also been considered [43]. Delayed triggering has been proposed by Parsons [45], and mentioned as possible by Gonzales-Huizar [46].

5.4 Cantilevered Beam Subjected to Light

Next we show one of the many results obtained with a cantilevered beam that was subjected to visible light. The test configuration was similar to the one in Fig. 15 but a lamp with a regular tungsten filament bulb was added to irradiate the beam from below. We compare three tests: one with only the ultrasound on, one with only the lamp on, and one with the ultrasound and the lamp both on.

The deflection of the beam with only the ultrasound on is seen on the left in Fig. 20. The beam was found to rise by 0.3 µm in 10 seconds. The two plots on the right in the Fig. 20 show the response of the cantilevered beam to only the lamp on: the beam sags by the same amount, 0.3 µm. The response when both the ultrasound and the lamp are on is shown in the middle of Fig. 20: The cantilevered end of the beam barely moves because the effects of the two activation sources cancel each other.

One important conclusion from this test is that the rock volume expands when subjected to the activation described. Note that we shone the light from below, causing the underside of the cantilevered beam to be illuminated. If the illumination leads to an expansion of a finite layer of rock on the underside, it would become understandable that the beam must rise. We know that ultrasound always softens the rock. However the ultrasound energy is always deposited into the entire rock volume. By contrast, illumination with visible light seems to affect only part of the rock, near the surface but enough to cause a mechanical response due to a partial volume expansion.

6 Free or Fixed Boundaries and New Questions

As seen by Fig. 10, the proposed model has fixed conditions. So far, most of the tests in the lab have been done during free boundary conditions. The boundary conditions can make a big difference, and this is therefore an important point. Any material expansion in volume in a fixed condition tends to increase the stress, which at first seems to be opposite the effect of an elastic modulus decrease for a free boundary condition. In order to clarify the details of the process, these opposed effects must be investigated more thoroughly. Work on this aspect started at the Blekinge Institute of Technology during the second half of 2012.

The shear and longitudinal elastic moduli are expected to behave differently due to (a) the mechanisms taking place in the material and (b) the boundary conditions being fixed or free in different directions.

A second important question to answer is whether there is a possibility that any identifiable specific signals may be produced during the recovery process, which seems to precede a rupture. We do not at the moment know the answer to this question. However, it can be argued that there should be a reciprocity between influence and effect. If there is an effect on other material parameters, i.e. if energy is absorbed, then the changes during the processes avalanching towards rupture might emit specific signals. It might thus be possible to identify material parameters that would change and identify a recognizable pattern. There may also be an asymmetry in the geometry and material properties, leading to transient abnormal stress-strain fields prior to the earthquake rupture. All these factors, however, would be disturbed due to the unstable conditions that develop when the elastic parameters change.

Should the model I have proposed prove correct, it imposes tight time constraints on quake prediction. Based on our studies of rock softening and recovery rates, I believe any pre-earthquake signal resulting from these processes could be detected no earlier than 20 hours before a major earthquake rupture.

A key question for earthquake prediction is whether the processes taking place during rock softening, differ sufficiently from the normal processes, to qualify as an identifiable and detectable pre-earthquake signal. For now this remains an open question. But we can say that without a period of recovery after the triggering (a delayed triggering), there would be no chance of identifying a point-of-no-return. In this case earthquakes would simply happen totally unannounced leaving the science community with nothing but statistical methods to rely on. Even if such statistical methods could be improved significantly by the future research, they would never provide certainty. The most promising conclusion from our studies to date is that, indeed, such a period of recovery appears to exist and that this period is linked in a deterministic way to the processes leading up to catastrophic rupture.

References

1. P. Varotsos, K. Alexopoulos, K. Nomicos, M. Lazaridou, Earthquake prediction and electric signals. Nature **322**, 120 (1986)
2. T.F. Freund, I.G. Kulachi, G. Cyr, J. Ling, M. Winnick, J. Tregloan-Reed, M.M. Freund, Air ionization at rock surfaces and pre-earthquake signals. J. Atmos. Sol.-Terr. Phys. (2009). doi:10.1016/j.jastp.2009.07.013
3. M. Hayakawa, S.S. Sazhin, Mid-latitude and plasmaspheric hiss: a review. Planet. Space Sci. **40**, 1325–1338 (1992)
4. V.A. Liperovsky, O.A. Pokhotelov, E.V. Liperovskaya, M. Parrot, C.-V. Meister, O.A. Alimov, Modification of sporadic E-layers caused by seismic activity. Surv. Geophys. **21**(5–6), 449–486 (2000)
5. S. Pulinets, K. Boyarchuk, *Ionospheric Precursors of Earthquakes* (Springer, Berlin, 2004)
6. S. Shalimov, M. Gokhberg, Litosphere-ionosphere coupling mechanism and its application to the earthquake in Iran on June 20, 1990. A review on ionospheric measurements and basic assumptions. Phys. Earth Planet. Inter. **105**(3–4), 211–218 (1998)
7. T. Dautermann, E. Calais, J. Haase, J. Garrison, Investigation of ionospheric electron content variations before earthquakes in southern California, 2003–2004. J. Geophys. Res. **112**, B02106 (2007). 20 pp. doi:10.1029/2006JB004447
8. T. Rabeh, M. Miranda, M. Hvozdara, Strong earthquakes associated with high amplitude daily geomagnetic variations. Natural Hazards Review **53**, 561–574 (2010). doi:10.1007/s11069009-9449-1
9. E.L. Afraimovich, E.I. Astafyeva, TEC anomalies, Local TEC changes prior to earthquakes or TEC response to solar and geomagnetic activity changes? Earth Planets Space **60**, 961–966 (2008)
10. J.T. Thomas, J.J. Love, M.J.S. Johnston, On the reported magnetic precursor of the 1989 Loma Prieta earthquake. Phys. Earth Planet. Inter. **173**, 207–215 (2009)
11. S.N. Postnikov, *Electrophysical and Electrochemical Phenomena in Friction, Cutting, and Lubrication* (Van Nostrand Reinhold Co., New York, 1978)
12. M.Y. Balbachan, I.S. Tomashevskaya, Change in Rock Strength as result of mechanical induction of charges. Dokl. Akad. Nauk SSSR **296**, 1085–1089 (1987)
13. K. Trachenko, Slow dynamics and stress relaxation in a liquid as an elastic medium. Phys. Rev. B **75**, 212201 (2007)
14. L. Cipelletti, L. Ramos, S. Manley, E. Pitard, D.A. Weitz, E.E. Pashkovski, M. Johansson, Universal non-diffusive slow dynamics in aging soft matter. Faraday Discuss. **123**, 237–251 (2003)
15. L. Ramos, L. Cipelletti, Ultraslow dynamics and stress relaxation in the aging of a soft glassy system. Phys. Rev. Lett. **87**(24), 245503 (2001)
16. K.C.E. Haller, C.M. Hedberg, Constant strain frequency sweep measurements on granite rock. Phys. Rev. Lett. **100**(6), 068501 (2008)
17. J.A. TenCate, E. Smith, R.A. Guyer, Universal slow dynamics in granular solids. Phys. Rev. Lett. **85**(5), 1020–1023 (2000)
18. P.A. Johnson, B. Zinszer, P.N.J. Rasolofosaon, Resonance and elastic nonlinear phenomena in rock. J. Geophys. Res. **101**, 11553–11564 (1996)
19. R.A. Guyer, P.A. Johnson, *Nonlinear Mesoscopic Elasticity: The Complex Behaviour of Rocks, Soils, Concrete* (Wiley-VCH, Weinheim, 2009)
20. R.A. Guyer, P.A. Johnson, Nonlinear mesoscopic elasticity: evidence for a new class of materials. Phys. Today **April**, 30–36 (1999)
21. J.A. TenCate, T.J. Shankland, Slow dynamics in the nonlinear response of Berea sandstone. Geophys. Res. Lett. **23**, 3019–3022 (1996)
22. K.C.E. Haller, C.M. Hedberg, Sound velocity dependence on strain for damaged steel, in *Nonlinear Acoustics—Fundamentals and Applications (ISNA 18)*, ed. by B.O. Enflo, C.M. Hedberg, L. Kari (Am. Inst. Phys, New York, 2008), pp. 271–274

23. E. Mfoumou, Low frequency acoustic excitation and laser sensing of vibration as a tool for remote characterization of thin sheets. Doctoral Dissertation Blekinge Institute of Technology, No. 2008:16, ISSN 1653-2090, ISBN 978-91-7295-155-6 (2008)

24. E. Mfoumou, K. Haller, C. Hedberg, S. Kao-Walter, Slow dynamics experiments on thin sheets, in *Proceedings of the Acustica 2008, Coimbra, Portugal, 20–22 October* 2008

25. J.A. TenCate, E. Smith, L.W. Byers, T.J. Shankland, Slow dynamics experiments in solids with nonlinear mesoscopic elasticity in nonlinear acoustics at the turn of the millennium: ISNA 15, in *AIP Conference Proceedings*, vol. 524 (2000), pp. 303–306

26. K.C.E. Haller, C.M. Hedberg, O.V. Rudenko, Slow variations of mechanical and electrical properties of dielectrics and nonlinear phenomena at ultrasonic irradiation. Acoust. Phys. **56**(5), 660–664 (2010)

27. S. Hough, *Predicting the Unpredictable—The Tumultuous Science of Earthquake Prediction* (Princeton University Press, Princeton, 2010)

28. Y. Gao, S. Crampin, Observations of stress relaxation before earthquakes. Geophys. J. Int. **157**, 578–582 (2004)

29. C.M. Hedberg, E. Johnson, S.A.K. Andersson, K.C.E. Haller, G. Kjell, S.-E. Hellbratt, Ultrasonic monitoring of a fiber reinforced plastic-steel composite beam during fatigue, in *Proceedings of 6th European Workshop on Structural Health Monitoring, Dresden, July 3–6*, 2012

30. Data from Advanced Composition Explorer satellite. http://www.srl.caltech.edu/ACE/ASC/level2/

31. R. Teisseyre, Generation of electric field in an earthquake preparation zone. Ann. Geophys. **15**, 297–304 (1997)

32. F. Freund, Charge generation and propagation in rocks. J. Geodyn. 545–572 (2002)

33. E. Odone, Tremblement de Terre 1904. Serie B. Catalogues. Publ. Strassburg (1907)

34. G. Duma, F. Freund, M. Lazarus, T. Rabeh, R. Dahlgren, Coupled ionospheric and telluric magnetic fields-seismotectonic relevance. EGU 2011, Vienna, 4–8 April

35. C.M. Hedberg, S.A.K. Andersson, K.C.E. Haller, Deflection dynamics of rock beam caused by ultrasound. Mech. Time-Depend. Mater. (2013). doi:10.1007/s11043012-9207-8

36. K.C.E. Haller, C.M. Hedberg, Method for monitoring slow dynamics recovery. Acoust. Phys. **58**(6), 713–717 (2012)

37. F. Pollitz, The 11 April 2012 $M = 8.6$ East Indian Ocean earthquake triggered large aftershocks worldwide. Public seminar at USGS Menlo Park, CA, Sept. 5, 2012

38. F.F. Pollitz, R.S. Stein, V. Sevilgen, R. Bürgmann, The 11 April 2012 east Indian Ocean earthquake triggered large aftershocks worldwide. Nature **490**, 11 (2012)

39. R.G. Roble, I. Tzur, *The Global Atmospheric-Electrical Circuit, The Earth's Electrical Environment* (National Academic Press, Washington, 1986)

40. F.T. Freund, Pre-earthquake signals. Part I: Deviatoric stresses turn rocks into a source of electric currents. Nat. Hazards Earth Syst. Sci. **7**, 1–7 (2007)

41. F.T. Freund, Pre-earthquake signals. Part II: Flow of battery currents in the crust. Nat. Hazards Earth Syst. Sci. **7**, 1–6 (2007)

42. H. Benioff, Earthquakes and rock creep: Part I: Characteristics of rocks and the origin of aftershocks. Phys. Rev. **144**(2), 469–477 (1951)

43. D.R. Shelly, Z. Peng, D. Hill, C. Aiken, Triggered creep as a possible mechanism for delayed dynamic triggering of tremor and earthquakes. Nat. Geosci. **4**, 384–388 (2011)

44. P.A. Johnson, X. Jia, Nonlinear dynamics, granular media and dynamic earthquake triggering. Nature **437**(6) (2005). doi:10.1038/nature04015

45. T. Parsons, A hypothesis for delayed dynamic earthquake triggering. Geophys. Res. Lett. **32**, L04302 (2005)

46. H. Gonzales-Huizar, A.A. Velasco, A. Peng, R.R. Castro, Remote triggering seismicity caused by the 2011, M9.0 Tohoku-Oki, Japan earthquake. Geophys. Res. Lett. **39**, L10302 (2012)

Steerable Nanobots for Diagnosis and Therapy

Anirudh Sharma, Yuechen Zhu, Madhukar Reddy, Allison Hubel, Ryan Cobian, Liwen Tan, and Bethanie Stadler

Abstract This paper reviews the synthesis of magnetic multilayerd nanowires that have a wide range of applications. Specifically of interest in this review are the applications of magnetic manipulation and separation of cells, which are important for potential cancer therapies. Compared to other magnetic nanoparticles in use today, the nanowires have the advantage of being ferromagnetic and also having high aspect ratios that enable barcoding. These nanobots are synthesized inside nanoporous oxide templates in large batches (10^{12} per square inch), and they can be composed of any magnetic metal, alloy, or multilayer that can be electroplated. Specific details for the electrochemistry of Galfenol deposition are given. Galfenol is an exciting new magnetostrictive material with durable mechanical properties. Next, a protocol is described for full removal of the growth contact prior to release of the nanobots from their oxide template. This mitigates aggregation which inhibits cellular uptake. Feasibility of manipulation and separation was shown using canine bone cancer (osteosarcoma) cells which internalized the nanobots, enabling magnetic cellular control. In addition, initial toxicity studies indicate that the nanobots are not cytotoxic. These studies merely scratch the surface of the potential use of nanobots for diagnosis and therapy in the near future.

A. Sharma · M. Reddy · B. Stadler (✉)
Electrical and Computer Engineering, University of Minnesota, Minneapolis, MN, USA
e-mail: stadler@umn.edu

Y. Zhu · A. Hubel
Mechanical Engineering, University of Minnesota, Minneapolis, MN, USA

A. Hubel
Biopreservation Core Resource, University of Minnesota, Minneapolis, MN, USA

R. Cobian
BH Electronics, Burnsville, MN, USA

L. Tan
Seagate Technologies, Bloomington, MN, USA

F. Freund, S. Langhoff (eds.), *Universe of Scales: From Nanotechnology to Cosmology*,
Springer Proceedings in Physics 150, DOI 10.1007/978-3-319-02207-9_18,
© Springer International Publishing Switzerland 2014

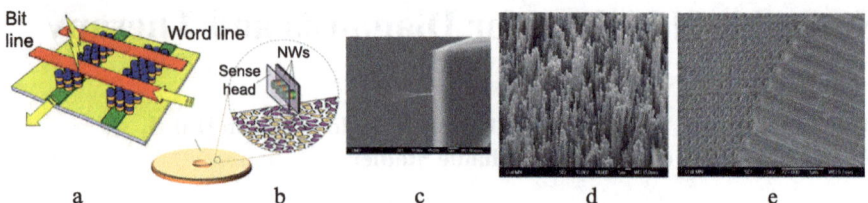

Fig. 1 (**a**) Schematics of potential random access memory using nanowires [1, 34]. (**b**) Hard drive read sensor arrays [2, 3]. (**c**) The topic of this paper, cilia or nanobots [24]. Note: the resonating cilia micrograph was taken at the University of Maryland by Alison Flatau's group. (**d**) Metallic nanowires [29] that have been fabricated inside the nanochannel "templates" shown in (**e**)

1 Introduction

The nanotechnology program that Minoru Freund had initiated at NASA Ames Research Center was from its outset designed to address a wide range of possible applications, including applications in medicine. For NASA the connection obviously came out of the need to explore new avenues in space medicine, specifically the need to provide diagnostic and therapeutic tools for future long-term travel into deep space.

Minoru Freund was an amazing person who had an enormous overview of what had been done in the past and deep insight into the material science, physics, and chemistry of nanomaterials. When he heard about magnetostrictive nanoparticles and that they could be made steerable by high frequency electromagnetic fields, his quick mind immediately caught on. We met in early 2011, when his brain tumor started to impair his motor control over the left side of his body, but he was as curious and as alert as ever.

Magnetic metallic nanowires are remarkably small rods of metal with diameters ranging from 5 nm to 200 nm and lengths that can extend up to 100's of microns. They can be synthesized in layered structures with great variability, and therefore they have promising potential in many technical applications. Some potential uses, illustrated in Fig. 1, include (a) bits in random access memory (RAM) [1], (b) read sensors in hard drives [2, 3] and (c) as cilia or flagella-like sensors and actuators [24]. Our group has grown "forests" of nanowires, as depicted in Fig. 1(d) [29], inside artificially created nanochannels (e) that run through the thickness of an aluminum oxide film or membrane.

To date, all read sensors and RAM bits start from vacuum-deposited thin films that are then patterned into devices via lithography followed by chemical etching [4, 5], similar to the schematic in Fig. 2(a). Sidewall damage from the etching steps is a major limitation in the performance of devices made by these techniques, once the dimensions move into the nanoscale because the surface/volume ratios increase dramatically. In short, the sidewall damage extends clear into the center of the device. In addition, vacuum-deposited films have classic columnar microstructures with many grain boundaries and voids, as shown in Fig. 2(b). When devices are made by standard lithography, their properties will vary as their size reaches that of the grains in the initial film. For example, in a 10 nm array of devices made by

Fig. 2 (**a**) Schematic of the popular top down approach to nanodevices. (**b**) Schematic of vacuum-synthesized thin films showing typical grain boundaries and voids [6]. (**c**) Schematic of lift-off lithography

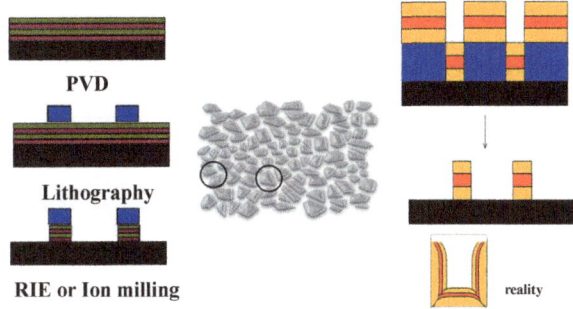

PVD

Lithography

RIE or Ion milling

reality

this standard process, some devices may be single crystal (one grain) while others have grain boundaries running through them in random locations as shown by the circles in Fig. 2(b). Another issue is that high-density arrays are difficult with standard lithography because the etch techniques typically involve ion milling (similar to atomic sand-blasting). The atoms removed from between the devices may re-deposit on the sidewalls of the devices themselves, which again causes properties to vary from the ideal. And, the same problem will occur that as devices are scaled to smaller volumes, these nonideal surfaces account for a higher and higher fraction of the device. In some cases, the lithography (top dots in Fig. 2(a)) is done before the device layers are deposited, so that the layers can be "lifted-off" by rinsing the resist, leaving the device behind, Fig. 2(c). Here, the layers more often deposit as shown in "reality", especially (again) as the features shrink into the nanoscale. So properties degrade quickly in nanodevices made by vacuum/lithography processes.

By contrast, our *in-situ* growth process produces devices without etching, completely mitigating sidewall damage. This paper focuses on nanowires to be used as artificial cilia and flagella. Since such nanowires can act as both, sensors and actuators, they have been referred to as nanobots [7]. Other groups producing cilia-like sensors in the millimeter scale are using polymers, which have the disadvantage they tend to degrade over short times [8–12].

Our nanowires are composed of layers of metals, usually magnetic/nonmagnetic stacks, which means they can have a wide range of tunable properties. One particularly appealing "smart" metal is magnetostrictive Galfenol (Fe$_{1-x}$Ga$_x$, $x = 0.1$–0.4). Figure 3 illustrates magnetostriction schematically, which is the strain response of a magnetic material to a magnetic field that changes its orientation. This phenomenon was discovered in bulk pure-metal magnets such as nickel ($\lambda_s = -40$ ppm), and recent materials such as Terfenol [13–15] and Galfenol [16–18] have been engineered that have giant magnetostrictions ($\lambda_s = 1600$ and 400 ppm, resp.). Galfenol is particularly interesting in that it has metallic mechanical properties such as ductility and strength, whereas the other giant magnetostrictive materials tend to be brittle and fracture readily. This is true even for metal alloys such as Terfenol. Therefore, despite having somewhat lower magnetostriction, Galfenol's durable mechanical properties are essential for cilia and nanobots because they need to be flexible and not break despite high aspect ratios, see Fig. 1(d). In addition, Galfenol is the only "smart" material that can be electroplated, unlike piezoelectrics and other

Fig. 3 Schematic illustrating magnetostriction, a physical phenomenon in which a material elongates due to the spin alignment in an applied field

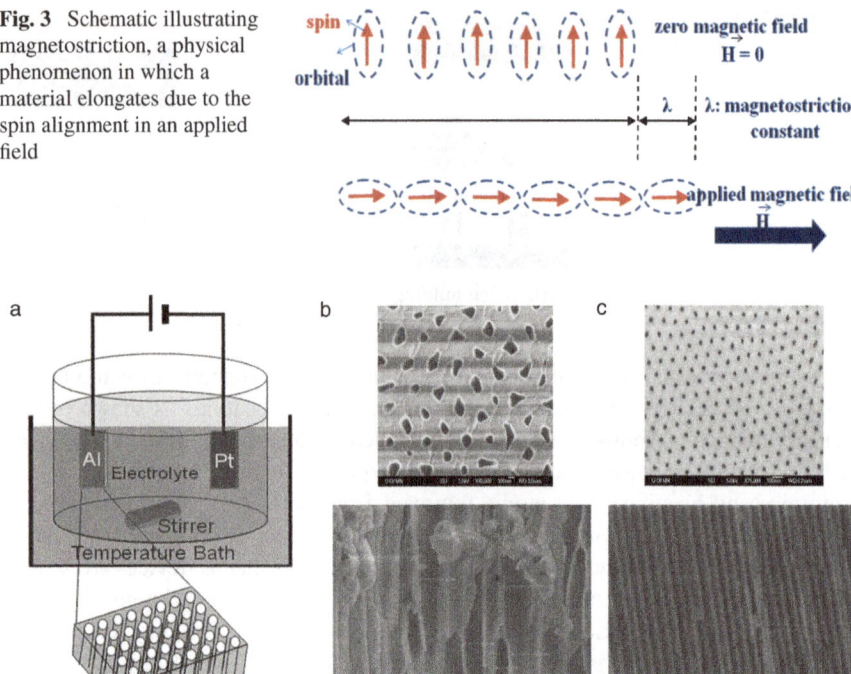

Fig. 4 (**a**) Schematic of the process of anodization. (**b**) SEM micrographs of commercial nanoporous filters by Whatman. (**c**) Nanoporous alumina made at the University of Minnesota [21]

magnetostrictives which either require oxide materials or rare earth metals [19, 30]. Our group has electroplated both films and nanowires of Galfenol, and recently we were the first group to measure magnetostriction in electroplated Galfenol [31, 32].

2 Nanowire Fabrication

In order to synthesize nanowires, nanochannels shown in Fig. 1(e) must first be made as templates into which the nanowires are grown by electrochemical deposition. These nanochannels are made by anodizing either bulk aluminum foil or aluminum films that have been deposited onto a substrate such as Si [33] as shown in Fig. 4(a). For nanobot applications, Al foils work best. But, for integrated devices such as read sensors and RAM, Al films on Si better enable subsequent integration with interconnects and the system platform.

When anodizing Al at the right temperature, with the right electrolyte, and the right voltage, nanochannels form in an oxide that grows on the aluminum metal. It has been shown that the interpore spacing (a) varies linearly with applied voltage (V) according to [20]

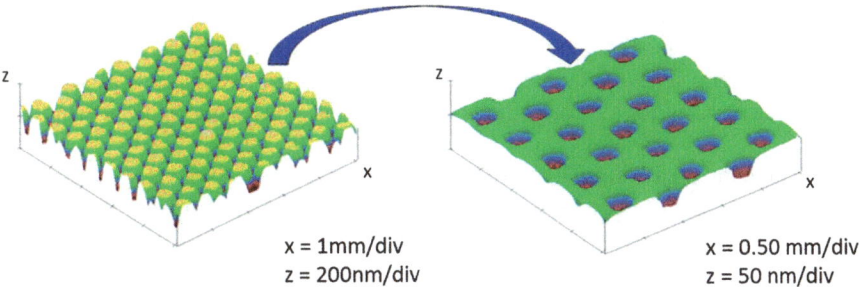

x = 1mm/div
z = 200nm/div

x = 0.50 mm/div
z = 50 nm/div

Fig. 5 Atomic force microscopy (*AFM*) images of Si3N4 nanoimprint stamp and the surface of aluminum that has been imprinted. Subsequent anodization of this aluminum will result in nanoporous alumina (that is, aluminum oxide) that is shown in Fig. 1(e) [2]

$$a \text{ (nm)} \sim -1.5 + 3 \text{ V (volts)} \tag{1}$$

for the entire range of a from 10 nm to 200 nm. Three electrolytes are usually used, and they each cover a different range of a, and therefore a different range of voltages according to (1). Sulfuric acid is used to grow nanopores with interpore spacings less than 30 nm, oxalic acid is used for $30 < a < 70$ nm, and phosphoric acid is used for $100 < a < 170$ nm. Larger spacings can be obtained by optimizing the concentration of the phosphoric acid. Typical pore diameters occupy from a third to a half of the interpore spacings. While reading literature on electroplated magnetic nanowires, it is important to note that some researchers use nanoporous templates that are sold as filters, Fig. 4(b). These nanopores are not exactly the columnar structures that are ideal for magnetic nanowires, Fig. 4(c) since magnetic flux will not be easily controlled.

To get channels that are homogeneous in size and uniformly spaced, such as shown in Fig. 1(e), nanostamps can be used to imprint the aluminum before anodization, Fig. 5. This patterned surface provides starting points for the nanochannels as the oxide begins to form, so the channels self-assemble inside the oxide as it continues to grow. At this point, our group can grow large-area nanoporous oxides in which billions and billions of nanochannels are all equal in diameters over a range of 10–200 nm depending on the anodization conditions.

To use these oxide films as templates, any remaining aluminum is etched away, and the bottoms of the pores are opened using phosphoric/chromic acid. The resulting nanoporous oxide membrane has a conductive coating sputtered onto one side, and a contact lead is soldered onto the coating. Both the lead and coating are then encased inside an acrylic polymer to insulate them from the electroplating bath.

The sample is then submerged in an electrochemical plating solution, and the only electrode material that is exposed to the bath is at the bottom of the nanochannels. Watts-type baths (containing metal sulfates) are usually used, and a voltage lower than the metal reduction potential is applied. The metal ions are reduced so that they plate sequentially up the channels to make nanowires. After nanowire growth, the contact is etched off of the back, and then the oxide is etched to release the nanobots. If the magnetic shape anisotropy is properly designed, we will have ferromagnetic nanobots. All of the other nano-particles that are currently used for magnetic separation or manipulation of cells are superparamagnetic, meaning they are not magnetic unless they are in a magnetic field. To give an idea of size, we can synthesize 2×10^{12} of our smallest nanobots in one square inch of template material.

Although the idea is simple, it can be a complicated process when trying to make nanobots of Galfenol because Ga is not an easy metal to electroplate. Rather, Ga oxides often form, followed by their precipitation, which depletes the bath of Ga all together. If oxides are avoided, it is common for nanowires of Galfenol to exhibit a bimodal distribution in length, Fig. 6(a) that is not observed with other metals, such as nickel, iron, gold or copper [19]. Unfortunately, the magnetic behavior of these nanowires is sharply dependent on their shapes. Therefore, the lengths must be controlled in order to control the nanobots. Using a rotating disk electrode, the diffusion boundary layer outside the nanochannels can be maintained to mitigate the parasitic growth of a few wires at the expense of the others, Fig. 6(b). Next, a copper seed layer can be grown at the base of the wires, and also pulsed deposition can be used to refresh the ion concentration inside the nanochannels, Fig. 6(c, d). These optimization techniques have been shown to drastically improve the uniformity in lengths from 78 % standard deviations to 3 % standard deviations [19].

3 Nanobot Applications

As can be seen from Fig. 1(c), these nanowires are mechanically flexible and strong. Further studies on their mechanical properties and on magnetic control of the multilayers have been done by our collaborators in Prof Alison Flatau's group at the University of Maryland [22–24]. After years of perfecting nanowire growth for magneto-electronic applications, such as the read heads and RAM, our group began to introduce nanowires to cellular assays more recently. The goal has been to determine the interactions between various cell types and these nanowires in order to provide novel diagnosis and therapy [25]. For diagnosis, we anticipate the ability to barcode cell types using segmented nanowires that have been functionalized and attached to specific cells. For therapy, nanowire-tagged cells can either be separated from specimens using an applied magnetic field, or they can be heated using an alternating applied magnetic field. The latter topic, called hyperthermia, is currently under study using superparamagnetic particles [26, 27]. It has been shown that if a tumor is heated, tumor cells will die 4 °F before healthy cells will die. This margin should be sufficient for magnetic nanomaterials to generate just enough heat due to hysteresis losses in alternating fields to only kill the cancer cells [28].

Initial studies have been performed using an osteosarcoma (OSCA-8) cell line that was derived from a canine tumor (Comprehensive Cancer Center, University of Minnesota). OSCA cells were cultured in Dulbecco's Modified Eagle Medium (DMEM, GIBCO) cell culture media supplemented with 10 % fetal bovine serum (FBS), Primocin (InvivoGen) and 4-(2-hydroxyethyl)-1-piperazineethanesulfonic

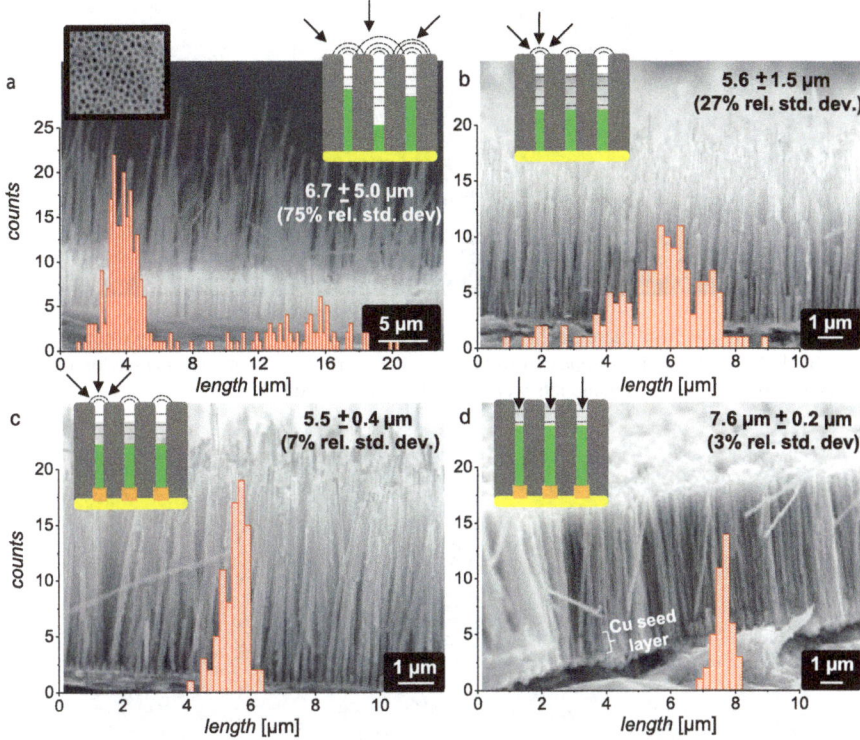

Fig. 6 Optimization of Fe-Ga nanowire growths. Statistical analysis of nanowire lengths is superimposed on the corresponding SEM image, and the schematics represent the diffusion profiles existing during growth in each case. (**a**) Bimodal distribution obtained when the solution was agitated using a magnetic stirrer, (**b**) after use of RDE-template at a rotation rate of 1800 rpm, (**c**) after use of Cu seed layer in addition to RDE-template and (**d**) use of pulse deposition in addition to Cu seed layer and RDE-template [19]

acid (HEPES, GIBCO). Cells were incubated at 37 °C and 5 % CO_2 atmosphere until highly confluent [25]. Nanobots suspended in phosphate buffered saline (PBS) were added to approximately 128 million cells in 6-well plates, after which they were incubated further for 24 hours. The cells were then trypsinized for resuspension and a 0.68 T magnet (SUPERMACs from Miltenyi) with 5.15 T/m gradient was used to separate of tagged cells. Figure 7 shows that osteosarcoma cells internalize Au/Ni/Au nanobots made according to the process described above. White arrows are pointing to the cell membranes and black arrows point to internalized nanobots. Initially, the nanowires were still aggregated due to incomplete etching of the growth contacts. Aggregates are not ideal, because the cells won't fully internalize the aggregates, as seen in Fig. 7b.

Etching protocols were established, including ion milling of the growth contact to ensure complete removal. Figure 8(a) shows an aggregate that still had the growth contact intact, also shown by the black film in the schematic. To solve this issue, the

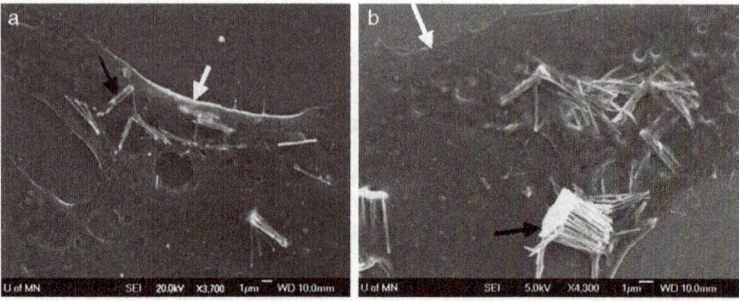

Fig. 7 Scanning electron micrographs of nanobots that have been internalized by osteosarcoma cells. (**a**) Individual nanobots appeared to be fully internalized inside membrane-enclosed compartments. (**b**) Aggregates of nanobots were not fully internalized [25]

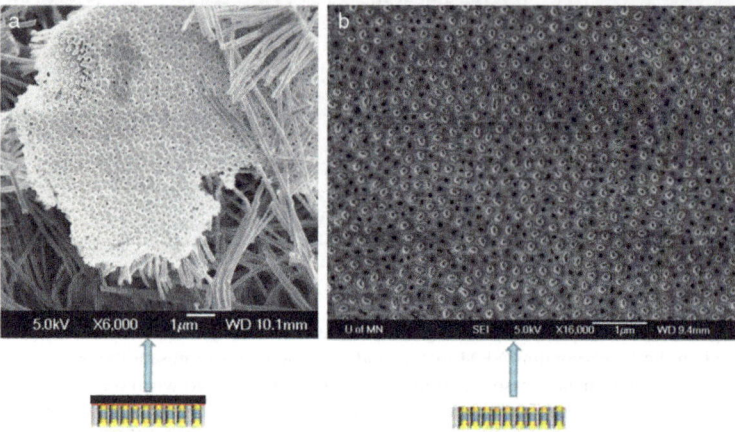

Fig. 8 (**a**) Incomplete etching of the nanobot growth contact resulted in aggregates. (**b**) Successful dispersions of nanobots required full etching of the growth contact prior to the dissolution of the oxide template [25]

growth contact was ion-milled after chemical etching to ensure that it was fully removed as seen in the SEM micrograph and schematic of Fig. 8(b). When the sample in Fig. 8(b) was placed into phosphoric/chromic acid to etch the aluminum oxide template, the nanobots were fully dispersed into the solution. In both cases, the nanobots were collected by a strong magnet for rinsing. As mentioned above, in the case of cell studies, the final suspension medium is phosphate-buffered solution (PBS) which is a popular cell culture medium.

Figure 9(a) shows transmission electron microscopy (TEM) analysis of osteosarcoma cells, which readily internalized the nanobots. This is a very common pathway for magnetic nanoparticle internalization.[1] Once the cells have internalized these

---

[1]See for example: http://www.magneticmicrosphere.com/meeting-ninth.

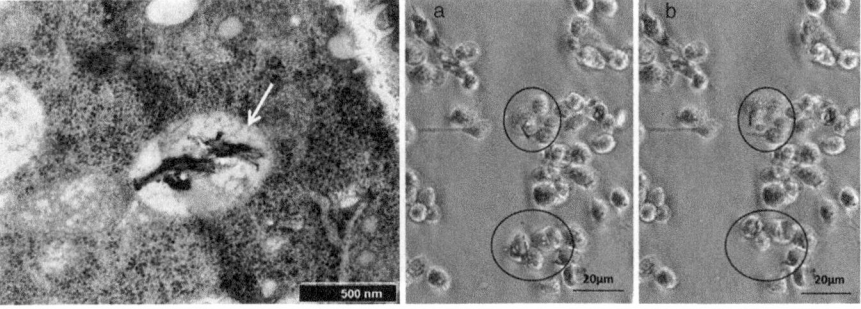

Fig. 9 (**a**) Transmission electron micrograph of nanowires inside intracellular vesicle after internalization by osteosarcoma cells. (**b**) Motion of cells was stimulated by external magnetic fields [25]

magnetic nanobots, they can be manipulated with an external magnetic field. One example of this is seen in Fig. 9(b, c). The cells were suspended in PBS after incubation with nanobots for 24 hours [25]. A NdFeB permanent magnet was rocked back and forth outside of the petri dish, and eventually the cells formed small aggregates due to the magnetic agitation.

Furthermore, magnetic separation of tagged cells was easily accomplished using these nanobots. Current methods of cell separation use commercially available magnetic microbeads, such as those sold by Miltenyi. Each cell is tagged using a cell-specific antibody that is chemically attached to the microbeads. The limitation of this technology is that the microbeads are all the same, so each cell type must be separated and identified one by one. This limitation can be overcome using the nanobots described here because their high aspect ratios and multilayer growth capabilities can produce an infinite number of barcodes and therefore magnetic properties. The ability to label populations of cells with unique barcoded nanowires would enable identification of multiple cell populations, also called multiplexing.

Initial studies of cell viability have also been done because it is important that the nanobots are not toxic to healthy cells. Figure 10 shows a micrograph in which 301 cells were found viable (stained with propidium iodide) and 2 cells out of 303 died (stained with acridine orange) after long incubations with nanobots. This was approximately the same time response of this cell line without nanobots. In addition, the cancer cells continued to proliferate at the normal rate, indicating that the nanobots were not toxic.

4 Conclusions

Magnetic manipulation and separation are important for potential therapies, but they can be accomplished with many types of magnetic nanoparticles. The nanowires shown here have the advantage of being ferromagnetic and also having high aspect

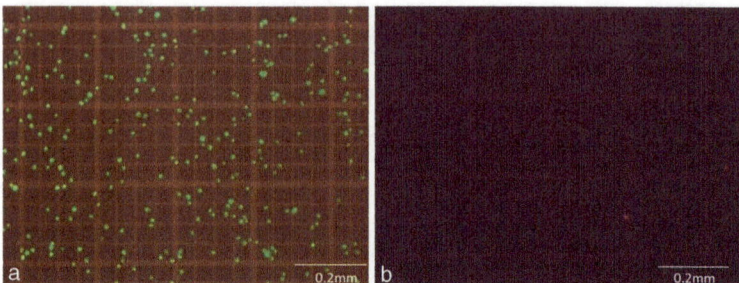

Fig. 10 Acridine orange and propidium iodide solution was mixed into cell assays after incubation with nanobots. (**a**) Live cells appear green and (**b**) dead cells appear *red* after staining. These results indicate low toxicity of nanobots [25]

ratios that enable barcoding. After nanobot synthesis, it was important to fully remove the growth contact prior to release of the nanobots from their oxide template. This enabled more nanobots to be available for cellular uptake. Initial studies with osteosarcoma cells indicate that cells will internalize nanobots for subsequent manipulation and separation from assays. In addition, initial toxicity studies indicate that the nanobots are not cytotoxic.

Acknowledgements   This work was primarily supported by the Grant in Aid and the Institute for Engineering in Medicine Programs of the University of Minnesota and supported partially by the MRSEC Program of the National Science Foundation under Award Numbers DMR-0819885. Parts of this work were carried out in the University of Minnesota Nanofabrication Center, Characterization Facility, and the University Imaging Center, which receive partial support from NSF through the NNIN program. This review covers over a decade of work by the following past and current group members: Patrick McGary, Liwen Tan, Xiaobo Huang, Rajneeta Basantkumar, Sang-Yeob Sung, Xiaoyuan Zi, Ratnanjali Khandwal, Vladimir Makarov, Andrew Block, Eliot Estrine, Matthew Hein, Ryan Cobian, and Neal Speetzen.

References

1. X. Huang, L. Tan, H. Cho, B.J.H. Stadler, Magnetoresistance and Spin transfer torque in electrodeposited Co/Cu multilayered nanowire arrays with small diameters. J. Appl. Phys. **103**, 07B504 (2009)
2. M. Maqableh, L. Tan, X. Huang, R. Cobian, G. Norby, R.H. Victora, B.J.H. Stadler, CPP GMR through nanowires (invited). IEEE Trans. Magn. **48**, 1–7 (2012). doi:10.1109/TMAG.2011.2174975
3. M. Maqableh, X. Huang, L. Tan, R.H. Victora, B.J.H. Stadler, Low resistivity 10 nm diameter magnetic sensors. Nano Lett. **12**, 4102–4109 (2012). doi:10.1021/nl301610z
4. Y.-S. Choi, H. Tsunematsu, S. Yamagata, H. Okuyama, Y. Nagamine, K. Tsunekawa, Stack structure of magnetic tunnel junction with MgO tunnel barrier prepared by oxidation methods: preferred grain growth promotion seed layers and Bi-layered pinned layer. Jpn. J. Appl. Phys. **48**, 120214 (2009)
5. J.R. Childress, M.J. Carey, S. Maat, N. Smith, R.E. Fontana, D. Druist, K. Carey, J.A. Katine, N. Robertson, T.D. Boone, M. Alex, J. Moore, C.H. Tsang, All-metal current-perpendicular-to-plane giant magnetoresistance sensors for narrow-track magnetic recording. IEEE Trans. Magn. **44**, 90–94 (2008)

6. R. Messier et al., J. Vac. Sci. Technol. A **2**(2), 500 (1984)
7. Private communication with Doug Rekenthaler
8. Y. Yang et al., in *Proceedings of the National Academy of Science*. Published online Nov. 28, 2006
9. B.P. Chagnaud et al., J. Neurosci. **V28**(17), 4479–4487 (2008)
10. G.J.M. Krijnen et al., Nanotechnology **17**, S84–S89 (2006)
11. N. Izadi et al., DTIP of MEMS & MOEMS, Stresa, Italy (2007)
12. Z. Fan et al., J. Micromech. Microeng. **12**, 655–661 (2002)
13. A.E. Clark, R. Abbundi, W.R. Gillmor, Magnetization and magnetic anisotropy of $TbFe_2$, $DyFe_2$, $Tb_{0.27}Dy_{0.73}Fe_2$, and $TmFe_2$. IEEE Trans. Magn. **MAG-14**, 542–544 (1978)
14. A.E. Clark, J.D. Verhoven, O.D. McMasters, E.D. Gibson, Magnetostriction in twinned [112] crystals of $Tb_{.27}Dy_{.73}Fe_2$. IEEE Trans. Magn. **MAG-22**, 973–975 (1986)
15. J.P. Teter, A.E. Clark, O.D. McMasters, Anisotropic magnetostriction in $Tb_{0.27}Dy_{0.73}Fe_{1.95}$. J. Appl. Phys. **61**, 3787 (1987)
16. A.E. Clark, J.B. Restorff, M. Wun-Fogle, T.A. Lograsso, D.L. Schlagel, IEEE Trans. Magn. **36**, 3238 (2000)
17. A.E. Clark, K.B. Hathaway, M. Wun-Fogle, J.B. Restorff, T.A. Lograsso, V.M. Keppens, G. Petculescu, R.A. Taylor, J. Appl. Phys. **93**, 8621–8623 (2003)
18. X. Zhao, D.G. Lord, J. Appl. Phys. **99**, 08M703 (2006)
19. K. Sai, M. Reddy, J.J. Park, S.-M. Na, M.M. Maqableh, A.B. Flatau, B.J.H. Stadler, Electrochemical synthesis of magnetostrictive Fe-Ga/Cu multilayered nanowire arrays with tailored magnetic response. Adv. Funct. Mater. **21**(24), 4677–4683 (2011)
20. K. Nielsch (Max Planck Institute, Halle Germany), Magnetic Nanowires, in *The Handbook of Magnetism and Advanced Magnetic Materials*, ed. by H. Kronmüller, S. Parkin (2007)
21. MSEE Thesis of Ryan Cobian, University of Minnesota (2004)
22. J.J. Park, M. Reddy, B.J.H. Stadler, A.B. Flatau, Hysteresis measurement of individual multilayered Fe-Ga/Cu nanowires using magnetic force microscopy. J. Appl. Phys. (2013, accepted)
23. J.J. Park, M. Reddy, C. Mudivarthi, P.R. Downey, B.J.H. Stadler, A.B. Flatau, Characterization of the magnetic properties of multilayer magnetostrictive Iron-Gallium nanowires. J. Appl. Phys. **107**, 09A954 (2010)
24. P.R. Downey, A.B. Flatau, P.D. McGary, B.J.H. Stadler, Effect of magnetic field on the mechanical properties of magnetostrictive iron-gallium nanowires. J. Appl. Phys. **103** (2008)
25. A. Sharma, Y. Zhu, S.S. Thor, F. Zhou, B. Stadler, A. Hubel, Magnetic barcode nanowires for osteosarcoma cell control, detection, and separation. IEEE Trans. Magn. **49**, 453–456 (2013)
26. A. Jordan, R. Scholz, P. Wust, H. Fahling, R. Felix, J. Magn. Magn. Mater. **201**, 413 (1999)
27. Y.H. Xu, J.M. Bai, J.P. Wang, J. Magn. Magn. Mater. **311**, 131 (2007)
28. R. Hergt, W. Andra, C.G. d'Ambly, I. Hilger, W.A. Kaiser, U. Richter, H.-G. Schmidt, IEEE Trans. Magn. **34**, 3745 (1998)
29. P.D. McGary, L. Tan, J. Zou, B.J.H. Stadler, P. Downey, A. Flatau, Magnetic nanowires for acoustic sensors (invited). J. Appl. Phys. **99**, 08B310 (2006). Also selected for publication in the Virtual Journal of Nanoscale Science & Technology **13** [18] (2006)
30. K. Sai Madhukar Reddy, E.C. Estrine, D.-H. Lim, W.H. Smyrl, B.J.H. Stadler, Controlled electrochemical deposition of magnetostrictive $Fe_{1-x}Ga_x$ alloys. Electrochem. Commun. **18**, 127–130 (2012). doi:10.1016/j.elecom.2012.02.039
31. E.C. Estrine, W.P. Robbins, B.J.H. Stadler, Electrodeposition and characterization of magnetostrictive Galfenol (FeGa) thin films for use in MEMS. J. Applied Physics (2013, accepted)
32. M. Hein, M. Maqableh, M. Delahunt, M. Tondra, A. Flatau, C. Shield, B. Stadler, Fabrication of BioInspired Inorganic Nanocilia Sensors. IEEE Trans. Magn. **49**, 191–194 (2013)
33. J. Zou, X. Qi, L. Tan, B.J.H. Stadler, Nanoporous silicon with long-range-order using imprinted anodic alumina etch masks. Appl. Phys. Lett. **89**, 093106 (2006)
34. Ph.D., Materials Science Thesis of Liwen Tan, University of Minnesota (2006)

Causal Factors for Brain Tumor and Targeted Strategies

Priya Ranjan Debata, Gina Marie Curcio, Sumit Mukherjee, and Probal Banerjee

Abstract Every five-year plan of each advanced country in the World includes major investments toward medical care. Consequently, vast improvements have taken place, bringing in precise robotic assistance in surgery and spectacular tools for the early detection of a large number of diseases. Advanced genomics and proteomics have ushered in promises for personalized medicine for cancer patients. Yet, the most advanced countries in the World still witness the highest proportion of age-adjusted incidence of brain cancer. Here we submit an overview of the reported etiology, genetics, and epigenetics that appear to be causal to cancers, especially for brain cancers. We discuss in some detail the use and usefulness of simple natural products such as curcumin to minimize the probability of developing cancer and to counteract existing cancers, even those as deadly as primary brain tumors. In this context we address the argument that brain cancers are more of a metabolic rather than a genetic disease and then discuss the acute need for new strategies for cancer therapy. Based on the findings from many laboratories including ours, we end this review advocating strongly for an effort to follow the example of Mother Nature and develop therapeutic strategies involving relatively safe food-derived anticancer agents.

1 Introduction

During the relatively short period of time that our group has been involved in research on brain tumors, we have been fortunate to meet some valiant fighters who have squarely faced the malady. Some of them were in the later stages of their lives, but a significant few were battling the disease during their earlier years, when their

P.R. Debata · G.M. Curcio · P. Banerjee (✉)
Department of Chemistry and the Center for Developmental Neuroscience, The City University of New York at The College of Staten Island, Staten Island, NY 10314, USA
e-mail: probal.banerjee@csi.cuny.edu

S. Mukherjee
CUNY Doctoral Programs in Biochemistry, The City University of New York at The College of Staten Island, Staten Island, NY 10314, USA

F. Freund, S. Langhoff (eds.), *Universe of Scales: From Nanotechnology to Cosmology*,   191
Springer Proceedings in Physics 150, DOI 10.1007/978-3-319-02207-9_19,
© Springer International Publishing Switzerland 2014

contemporaries were busy critically assessing many prosaic aspects of their appearance, social importance, employment, relationships, etc. The symposium, "A Universe of Scales: from Nanotechnology to Cosmology", which celebrated the life of a brilliant scientist, Dr. Minoru Freund, and the memory of his strong interest in research on food-derived natural agents have re-invigorated our effort. We are convinced that new, safe and effective therapies will eventually evolve from such natural agents that have always protected a large section of the human population from developing cancer. This review attempts to summarize some knowledge acquired by a large number of highly established teams performing state-of-the-art cancer research and then introduces the concept of potentiating natural agents to combat the so-far unvanquished malady of brain cancer.

2 Etiology, Symptoms, and Available Treatments for Brain Cancer

Cancer is an age-old disease and of diverse etiology. A normal cell becomes cancerous because of genetic and epigenetic modification in DNA and histones, differential expression of genes, and exposure to different forms of carcinogens, which eventually lead to the cancer-associated metabolic disorders. The process of tumorogenesis starts from aberrant division one cell and finally diverges into a group of clones with distinct molecular signatures. For multiple reasons, cancer treatments using a single drug have led to only marginal success, so patient-specific personalized treatment options are being considered now [1]. Aside from some common processes, such as increased cell proliferation, angiogenesis, and metastasis and a few proteins involved in these processes, overwhelmingly diverse mechanisms regulating the numerous signaling pathways involved in carcinogenesis have clearly established that cancer is a group of diseases that require diverse treatments.

Brain tumors are aggressive neoplasms afflicting both children and adults. The major risk factors associated with brain tumors include radiation exposure [2, 3], inherited abnormal genes from a parents [4], exposure to inorganic lead, non-arsenic insecticides, work in petroleum refining [5], occupational exposure to carbon tetrachloride [6], and viral infections [7]. Other risk factors include exposure to pesticides, first-hand or second-hand smoking, family history of cancer, antihistamine intake [8], maternal diet of cured meat [9], cured meats in adults and dietary cured meat in pregnancy and perinatal application of n-nitroso compounds [10]. The levels of nitroso compounds are suppressed by vitamin C and E, which are among the protective factors [10].

Although such risk factors play a central role in the etiology of cancer, it is important to note at the outset that cancer seems to involve a reversal of the cell's metabolic machinary to a primitive state when organisms required much less oxygen than current aerobic organisms [11]. Recent studies have demonstrated that hypoxic conditions can really suppress important proteins like bone morphogenic protein (BMP), which trigger maturation of cells into specific undividing or slowly dividing brain cells [12]. This viscious cycle of a change in the metabolic framework and the consequent suppression of cell maturation yields a perpetually-dividing cancer cell. Fortunately, a human body harbors a powerful immune system that promptly eliminates these rogue cells. Genetic and metabolic changes prompting such transformations will be discussed in this review.

A tumor in the brain increases intracranial pressure, leading to symptoms, which include headache, nausea, vomiting, blurred vision, imbalance, changes in personality and behavior, seizures, drowsiness or even coma [13]. There are two groups of brain tumors with respect of their origin. Primary brain tumors originate in the brain and metastatic brain tumors initially develop perpherally and then metastasize into the brain. Malignant brain tumors, particularly glioblastomas, are primary brain tumors that are extraordinarily difficult to treat [14]. Ironically, the developed industrial countries with the best medical care have the highest age-adjusted incidence rate of brain tumor [10]. The prognosis of brain tumor is very poor because of therapeutic resistance and recurrence of tumor after surgical resection. The recurrence is greatly due to a population of cancer cells called cancer stem cells that are treatment resistant and capable of proliferation [15]. The available treatment includes surgical removal of the tumor, followed by chemo and radiation therapies. They are painful and in many cases they kill normal cells and trigger angiogenesis (new blood vessel formation to nourish the tumor).

Many therapeutic agents that have potential to kill cancer cells are not effective in brain tumor therapy because they cannot cross the blood-brain barrier (BBB). Temozolomide (TMZ; a.k.a. Temodar), an alkylating agent that can cross the BBB, is used in a standard chemotherapy regimen [16, 17]. Radiotherapy followed by adjuvant TMZ treatment increases patient survival [18], but such treatments are associated with neurotoxicity and severe cognitive disturbances [19]. Biological treatments include a humanized monoclonal antibody Bevacizumab (Avastin), which is meant to block angiogenesis [20]. Despite the multitude of available treatment options, the average survival of glioblastoma patients is only 15 months, and therefore, there is a strong need for new therapeutic options for brain tumor.

One major reason for this limited success lies in the fact that the chemotherapeutic drugs also kill normal cells. This leads to severe side effects, which make the patient's life agonizing. Development of a targeted delivery approach has been under investigation as a relatively new approach. It involves selective delivery of cancer drugs to tumor cells. Tumor cells over-express receptors or specific proteins, which occur at low levels in normal cells. In many cases these cancer-cell-specific proteins functionally participate in pathways that are involved in the oncogenic process in gliomas. These cascades often involve the epidermal growth factor receptor (EGFR), platelet-derived growth factor receptor (PDGFR), vascular endothelial growth factor receptor (VEGFR), phosphoinositide 3-kinase (PI3K), and the signaling members of the mammalian target of rapamycin (mTOR) and the mitogen-activated protein kinase (MAPK) pathway [21]. High expression of these proteins is most often observed in glioma cells. Prominin-1 (CD133) a stem cell marker, is now extensively used as a surface marker to identify and isolate brain tumor stem cells (BTSCs) in malignant brain tumors [22]. The brain tumor stem cells can also differentiate into endothelial cells and trigger angiogenesis [17]. Furthermore, cancer cells derive a special advantage for survival under adverse conditions by remodelling many metabolic pathways and rendering multiple regulatory processes ineffective. This concept will be discussed in greater details later in sections V and VI.

3 Genetic Changes in Brain Cancer

Many cancer researchers argue that each cancer is strongly linked to the genetic make-up of subject and as such, a personalized cancer treatment could probably be designed for the patient. However, like many other strategies, gene-based therapeutic strategies have yielded little success over several decades and brain cancer has in most cases evaded such therapy. Screening genes for proteins such as p53 and XRCC have yielded mixed results. As for mutations in p53, one study shows an associated increase in risk of brain cancer [23], while a second report reveals no link to the disease [24]. Other groups report that p53 is mutated in 40 % of astrocyte tumors, most commonly in gliomas [25]. A polymorphism in the p53-encoding gene *TP53* has also been reported in glioblastoma [26]. Mutations in *TP53* and *PTEN* are also detected in glioblastoma [27]. Thus, although no direct evidence has been reported, a likely association of p53 metations to brain cancer remains as a likely possibility. A mutation in the pro-apoptotic protein Caspase 8 has been reported as a risk factor for glioma [28]. Inverse association of polymorphism in genes linked to inflammation and asthma with glioblastoma has also been reported [29]. In astrocyte tumors, mutations in *CDKN2*, which encodes p16 has been reported. P16 functions to inhibit CDK4, which when complexed with cyclin D1, phosphorylates and inhibits Rb (a tumor suppressor). Consistent with this, in glioblastoma multiforme, mutations in both *Rb* and *p*16 have been observed, although both genes are rarely activated in the same cell. Intriguingly, in glioblastoma multiforme (GBM) cells where either of the above mutations is not observed, CDK4 over-expression has also been reported [30]. Deletion of a part of chromosome 10 has been reported in some cancers [31]. This chromosome harbors many tumor suppressor genes, namely *PTEN*, and loss of function of PTEN has been implicated in more malignant, high grade tumors [32]. Multiple *CDK/CYCLIND* genes are amplified in medulloblastoma and supratentorial primitive neuroectodermal brain tumor. Among the aberrant genes that are linked to brain tumors, the *BRAF* gene is frequently duplicated [33]. Tandem duplication at 7q34, leading to a fusion between *KIAA*1549 and *BRAF* (f-BRAF) with increased BRAF *activity* is also observed in pilocytic astrocytomas [34, 35]. Two *EGFR* gene variants (rs17172430 and rs11979158) are associated with homozygous deletion at the *CDKN2A/B* locus [36]. Two single nucleotide polymorphisms (SNPs) at the promoter region, -216G/T and -191C/A, and a polymorphic (CA)(n) microsatellite sequence in intron 1 increase the risk of glioma [37]. Heterogeneity of subcellular localization of p53 protein in human glioblastomas: A study in Korean population reports a *AICDA/CASP14* polymorphism linked to childhood brain tumor [38]. Some meningioma tumors are associated with gene mutations in TNF receptor-associated factor 7 (TRAF7), Krupple-like factor 4 (KLF4), v-AKT murine thymoma viral oncogene homolog 1 (AKT1), and the Smoothened, Frizzled family receptor (SMO) [39].

4 Epigenetic Modifications in Brain Tumor

Expression of specific cancer-linked genes is regulated by epigenetic mechanisms such as DNA methylation, histone modifications and noncoding RNAs, and abnormalities in these processes have been found in multiple cancer types. DNA methylation can cause downregulation of gene expression and gene silencing. The CpG islands in gene promoters are rich in GC dinucleotide and are targets for DNA methylation. Epigenetic modifications have been reported in several cases of brain tumors. These include hypermethylation of proximal promoter of transglutaminase 2 gene [40], glial fibrilary acidic protein (GFAP) gene promoter [41]. *High frequency methylation of retinoic acid receptor β (RARβ) and O-6-methylguanine-DNA methyltransferase (MGMT)* in primary glioblastoma [42]. WNT inhibitory factor-1 (WIF1) promoter hypermethylation in astrocytoma [43]. CpG island methylation of the urea cycle enzymes argininosuccinate synthetase (ASS) and argininosuccinate lyase (ASL) is reported in GBM [44]. Similar methylation is also observed in *GATA4* and *DcR1* gene promoters [45]. The promoter of the gene for the inhibitor of DNA binding/differentiation transcription factor 4 (Id4) was methylated in 37 % cases of GBMs [46]. CpG island hypermethylation was observed in mutant isocitrate dehydrogenase 1 (IDH1) in gliomas [47] and the promoter for the tumor necrosis factor receptor superfamily member *11A* gene (*TNFRSF11A*) [48]. *MGMT*, *GATA6*, and *CASP8* genes methylation is also observed in glioblastoma [49]. Hypermethylation of the 5' region and untranslated first exon of the secretory granule neuroendocrine protein 1 gene (*SGNE1/7B2*) occurs in gliomas [50]. Similar DNA methylation silences the nonsteroidal anti-inflammatory drug-activated gene (*NAG-1/GDF15*) in glioma cell lines [51].

Histone acetylation and deacetylation are important events in gene regulation. Acetylation at a lysine amino acid residue of histone is catalyzed by the enzyme histone acetyltransfereases (HAT). Similarly, removal of an acetyl group from a lysine residue is catalyzed by histone deacetylases (HDACs). Both processes cause chromatin remodelling and regulate gene expression. Acetylation neutralizes the positive charge on lysine residue and impairs binding of histone to negatively charge DNA, thereby facilitating gene expression. In contrast, histone deacetylation restores the positive charge on lysine residue, thus causing binding of histone protein to negatively charged DNA, which is generally linked to transcriptional repression.

The HDAC inhibitor trichostatin A has been shown to inhibit glioma cell proliferation [52]. Similarly another HDAC inhibitor, MS275, sensitizes glioblastoma cells for chemotherapy-induced apoptosis [53, 54]. MS-275, valproic acid or SAHA, provide a novel strategy for sensitization of medulloblastoma to DNA-damaging drugs such as Doxorubicin, VP16, and Cisplatin by promoting p53-dependent mitochondrial apoptosis [55].

Intriguingly, such chemotherapeutic strategies hold brain cancer in check for some time, before it overpowers such treatments. Therefore, it is important to seriously reconsider other aspects of brain cancer. The most general feature of a cancer cell is its oxygen-insensitive, 200-fold higher rate of glycolysis in the cytoplasm, leading to fermention-like formation of lactic acid even in the presence of high oxygen tension. This is probably due to adaptation of cancer cells to the frequent hypoxic conditions observed inside fast-growing tumors [56]. Thus, cancers may constitute a metabolic rather than genetic disease.

5 Is Cancer Primarily a Metabolic Disease?

Dysregulated and reprogrammed metabolism is a hallmark of cancer, including brain cancers [57–59]. Cancer cell lines have a higher rate of glycolysis compared to oxidative phosphorylation [60]. Additionally, GBM cells have enhanced rates of both glycolysis and glutaminolysis [57]. Intriguingly, in human glioblastoma orthotopic tumors derived from independent GBMs, the anaplerotic flux for glutamine formation is not exactly balanced by its cataplerosis in brain tumor cells, which leads to the accumulation of a large glutamine pool and minimal glutaminolysis [57, 61]. The resultant metabolic anomaly is the accumulation of a large glutamine pool within the tumor [61]. In contrast to normal neurons and glial cells in brain, malignant brain tumors cannot utilize ketone bodies for energy and are heavily dependent on glucose and glutamine for their energy needs [62–64]. Analysis of *in vivo* [U-^{13}C]glucose metabolism in metastatic high-grade gliomas in human subjects show that less than 50 % of the acetyl-CoA pool is derived from blood-borne glucose, which suggests that other substrates also contribute in a major way to tumor metabolism. Additionally, lactate production is sharply increased in metastatic GBMs [65].

6 The Need for a New Strategy for the Treatment of Cancer

In cancer treatment, increasing use of targeted anticancer agents that inhibit tyrosine kinase signaling (monoclonal antibodies or tyrosine kinase inhibitors) (trastuzumab, sunitinib) have improved the survival of patients with malignancies, but cardiotoxicity, including heart failure, left ventricular dysfunction, hypertension, myocardial infarction, and thromboembolism, has accompanied such treatment. Similarly, other anticancer agents currently used for targeting, such as maytansinoids, calicheamycin, or auristatins belong to this category in which the untargeted form of each drug is toxic toward normal cells [66]. Recently the only payload-containing anticancer drug, mylotarg (calicheamycin-based), was withdrawn because of serious side effects. Furthermore, neurotoxicity from mechanical trauma (surgery), radiotherapy, and chemotherapy can increase extracellular concentrations of glutamate, which is converted into glutamine that particularly nourishes the glioblastoma cells

[67]. In addition to such difficulties, a wide range of signaling molecules are dysregulated in cancer and such targeting against specific signaling molecules is therefore unlikely to be effective as a long-term therapeutic strategy. This view is also held by other scientists, and it has been proposed that targeting multiple hallmarks of cancer could enable us to overcome the typical drug resistance acquired by many cancer cells [68]. This emphasizes the need for an alternative, safe strategy for cancer treatment.

(i) *A glimmer of hope, but challenges galore*: Brighter prospects have been revealed by epidemiological research, which has shown that dietary factors may play a highly important role in cancer prevention [10, 69, 70]. In fact, the efficacy of food components like resveratrol (from red grapes), epicatechin gallate (from green tea), and curcumin (a component of the spice turmeric) in killing a wide variety of cancer cells has been established in numerous studies. Although these agents harbor strong prophylactic activity against cancer formation [71, 72], they have been uniformly ineffective in eliminating established tumors.

Curcumin, a polyphenolic compound isolated from Curcuma longa (Turmeric) is widely used in traditional Indian Ayurvedic medicine [73]. Curcumin has the ability to fight the cancer cells by modulating the hallmarks of cancer. Curcumin causes G2/M phase arrest and inhibit JAK1,2/STAT3 signaling pathways in a glioma model [74]. It promotes differentiation and autophagy [75], inhibits proliferation, invasion and metastasis [76], induces apoptosis [77], tumor angiogenesis [78], blocks brain tumor formation [72], suppresses anti-apoptotic signals [79], inhibits telomerase activity [80]. In gliomas, induced expression of receptor tyrosine kinase activities (e.g. EGFR, PDGFR or VEGFR) leads to cell survival and proliferation. Curcumin has been shown to suppress the expression of these proteins in various types of cancer including glioma [81, 82]. Intriguingly, curcumin has been shown to inhibit HDAC activity in medulloblastoma cells [83], although curcumin-mediated hypoacetylation of histones has also been reported [84–86].

(ii) *Why would curcumin kill cancer cells without harming normal cells?* First, curcumin interacts with thioredoxin reductase (TrxR), which is overexpressed in tumor cells. This interaction promotes alkylation of TrxR at its catalytic site, thereby converting it into an NADPH oxidase, which in turn results in increased production of harmful reactive oxygen species in the tumor cells [87]. Secondly, glutathione levels in tumor cells are generally lower than in normal cells. Curcumin evokes further inhibition of glutathione and the ensuing superoxide formation promotes apoptosis (programmed cell death) in the tumor cells [88]. As expected, curcumin causes no superoxide injury to normal cells, which contain higher levels of glutathione. Finally, the protective transcription factor NF-kB, which is constitutively activated in most cancer cells, is strongly inhibited by curcumin [89], which triggers apoptosis of these cells.

(iii) *Fresh challenges and new strategies*: As is true also for resveratrol, the beneficial ingredient in red wine, the therapeutic potential of curcumin is seriously hindered due to the combination of low intestinal absorption, rapid metabolism in the body, and low solubility in water. This results in an overall low bioavailability of curcumin in the bloodstream [72]. In order to improve its bioavailability alternative ways of delivery are under investigation. For example, nanoparticle-based drug

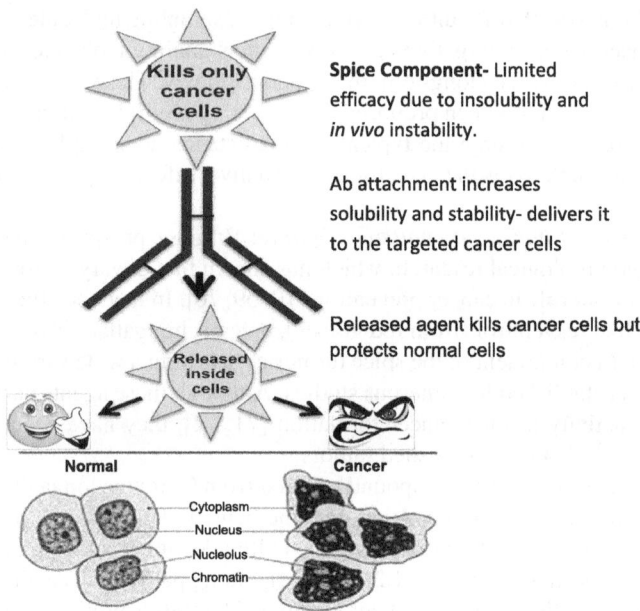

Spice Component- Limited efficacy due to insolubility and *in vivo* instability.

Ab attachment increases solubility and stability- delivers it to the targeted cancer cells

Released agent kills cancer cells but protects normal cells

Fig. 1 The safety of approach and uniqueness of our strategy. It should be noted that our strategy is to deliver the immuno-nutraceutical not systemically, but by direct intracranial infusion through a cannula. This is similar to a human arrangement termed "shunt"

delivery has been investigated and preclinical studies have shown promising results. These include magnetic nanoparticle [90], and methoxy polyethylene glycol-poly(caprolactone) nanoparticles [91]. Such strategies have greatly improved the stability and bioavailability of curcumin, but we still need to design therapeutic ways to have curcumin reach the brain tumor cells at high enough concentrations to have an effect on brain cancer. We believe that this can be achieved mainly by antibody-mediated targeting of curcumin in a releasable form to the brain tumor cells (Fig. 1). Our studies have demonstrated that targeting of curcumin to cancer cells by linking it in a releaseable form to an antibody again a glioblastoma surface marker CD68 converts this innocuous food component into a potent agent that can eliminate established glioblastoma cells *in vitro* and *in vivo* [82, 92].

As shown in Fig. 2, treatment of cultured mouse glioblastoma cells GL261 with this targeted version of curcumin elicited an increase in fluorescence in the cells, indicating release of curcumin from the curcumin-CD68 antibody adduct by the action of intracellular esterases (Fig. 2a–c). Within two hours of adduct treatment extensive blebbing and then complete decimation of cells were observed (Fig. 2f–h). A similar adduct treatment of GL261 cells for 24 hours caused a dramatic increase in the activity of the pro-apoptotic enzymes caspase-3 and 7 (Fig. 2e).

(iv) *In a clinical setting, what would be the best mode of intracranial delivery of the curcumin adduct*? The current mainstream brain tumor therapy consists of tumor resection followed by radiation and chemotherapy. The most common chemothera-

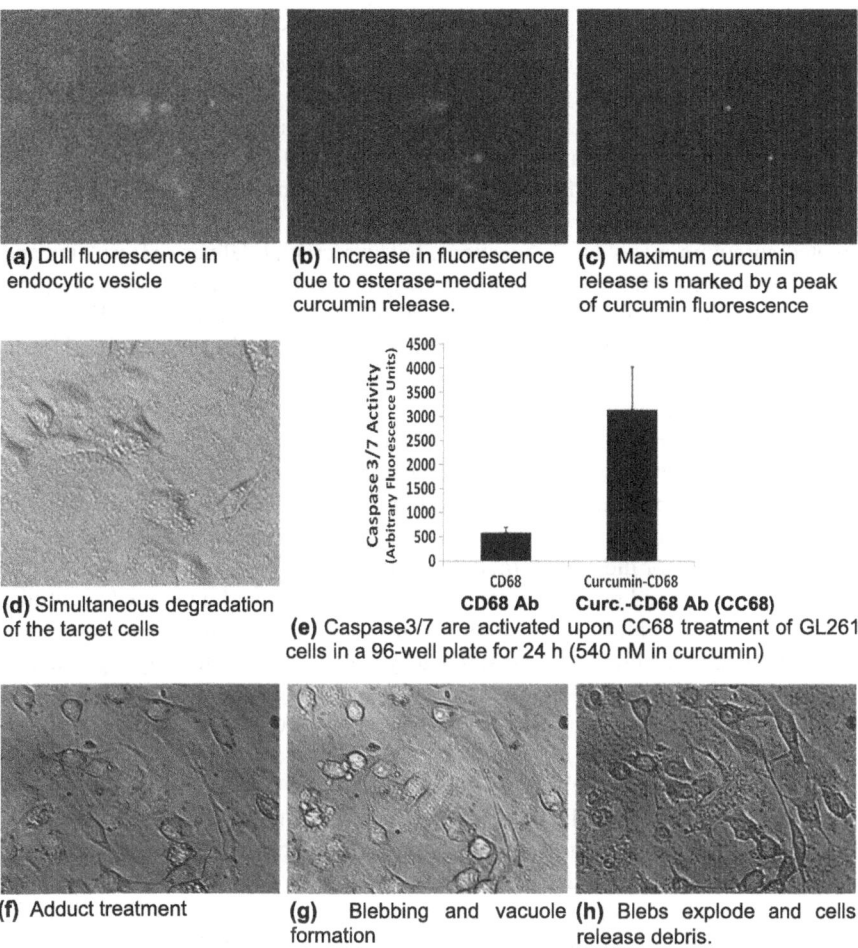

(a) Dull fluorescence in endocytic vesicle

(b) Increase in fluorescence due to esterase-mediated curcumin release.

(c) Maximum curcumin release is marked by a peak of curcumin fluorescence

(d) Simultaneous degradation of the target cells

(e) Caspase3/7 are activated upon CC68 treatment of GL261 cells in a 96-well plate for 24 h (540 nM in curcumin)

(f) Adduct treatment

(g) Blebbing and vacuole formation

(h) Blebs explode and cells release debris.

Fig. 2 Time course of glioblastoma destruction. Mouse glioblastoma GL261 cells in a 96-well plate were treated with 540 nM CD68 antibody-linked curcumin (CC68) followed by time-lapsed imaging using alternating epifluorescence and brightfield imaging for two hours. (**a–c**) Curcumin release (from the Curcumin-CD68 Ab adduct) in the target cells is marked by an increase in curcumin fluorescence at 520 nm. (**d**) A brightfield image corresponding to (**c**) shows degradation of cells. (**e**) Treatment of GL261 cells for 24 hours with CD8 antibody alone or CC68 causes a dramatic increase in caspase-3/7 activity the CC68-treated wells. (**f–h**) After treating the GL261 cells as in (**a**), time-lapsed images were acquired only with phase-contrast imaging for two hours, during which CC68 caused blebbing and destruction of the GL261 cells

peutic agent used for brain tumor therapy is temador, which crosses the blood-brain barrier but also produces agonizing side effects [19, 93]. Additionally, as mentioned earlier, radiotherapy and chemotherapy can increase extracellular glutamate, which is converted into glutamine and consumed as fuel by brain tumor cells, in particular glioblastoma [94]. Even though intracranial surgery is the first step taken in brain tumor therapy and intracranial application of the monoclonal antibody Avastin has been performed (see below), the use of systemically-delivered agents like Temozolomide have, unfortunately, created the strong notion in the medical community that peripheral delivery is the only application route for anti-cancer agents.

The human-compatible arrangement for intracranial delivery is a "shunt" that is often used to drain fluid from the brain. The same arrangement is also used to deliver drugs directly into the brain tumor. At the outset, the intracranial delivery of targeted curcumin raises many questions, but intracranial delivery of proteins and peptides has been used as a therapeutic strategy for neurodegenerative diseases [95]. This study [91] reports intracranial delivery of the trophic factor GDNF into the dorsal putamen, which resulted in improved motor function in advanced Parkinson disease patients. In cancer therapy, the popular brain cancer drug Avastin, which is a monoclonal antibody, has been delivered using microcatheters into the blood vessels inside the brain, close to the site of the tumor in brain tumor patients [96]. Such studies also provide additional information that proteins, which are expected to be blocked from brain entry by the blood-brain-barrier could be made to enter the brain by the use of mannitol, a sugar, that transiently opens the blood-brain barrier and has been used for the peripheral delivery of Avastin.

The evidence-based possibility that intracranial delivery of a targeted agent like antibody-linked curcumin could expunge cancer cells, eliminate inflammation, and ensure prolonged health cannot be summarily cast aside. We strongly argue that it is time to think outside the set notion and apply the targeted curcumin therapy intracranially to rescue brain tumor patients. As mentioned earlier, targeting multiple hallmarks of cancer could give us a better chance to overcome the drug resistance acquired by many cancer cells [68]. Since natural food components like curcumin target multiple hallmarks of cancer, appropriate targeting of such agents is likely provide a promising, new-age therapy for cancer. For Dr. Minoru Freund these new opportunities to attack and destroy a brain tumor as devastating as GBM came too late. However, all through his 2 1/2 year long battle, until the last months of his life, he was deeply involved in discussions how to bring to bear the power of nanotechnology for developing effective and selective drug delivery systems.

Acknowledgements   A fellowship support for Sumit Mukherjee from the CUNY Graduate Center is gratefully acknowledged here.

References

1. J.E. Wolff et al., Preliminary experience with personalized and targeted therapy for pediatric brain tumors. Pediatr. Blood Cancer **59**(1), 27–33 (2012)
2. E.C. Peterson et al., Radiation-induced complications in endovascular neurosurgery: incidence of skin effects and the feasibility of estimating risk of future tumor formation. Neurosurgery (2012)
3. M.Z. Braganza et al., Ionizing radiation and the risk of brain and central nervous system tumors: a systematic review. Neuro-Oncol. **14**(11), 1316–1324 (2012)
4. K. Hemminki et al., Familial risks in nervous-system tumours: a histology-specific analysis from Sweden and Norway. Lancet Oncol. **10**(5), 481–488 (2009)

5. T. Brown et al., Occupational cancer in Britain. Remaining cancer sites: brain, bone, soft tissue sarcoma and thyroid. Br. J. Cancer **107**(Suppl 1), S85–S91 (2012)
6. J.S. Nelson et al., Potential risk factors for incident glioblastoma multiforme: the Honolulu heart program and Honolulu-Asia aging study. J. Neurooncol. **109**(2), 315–321 (2012)
7. K. Alibek, A. Kakpenova, Y. Baiken, Role of infectious agents in the carcinogenesis of brain and head and neck cancers. Infect. Agents Cancer **8**(1), 7 (2013)
8. S. Cordier et al., Incidence and risk factors for childhood brain tumors in the Ile de France. Int. J. Cancer **59**(6), 776–782 (1994)
9. J.M. Pogoda et al., An international case-control study of maternal diet during pregnancy and childhood brain tumor risk: a histology-specific analysis by food group. Ann. Epidemiol. **19**(3), 148–160 (2009)
10. H. Ohgaki, P. Kleihues, Epidemiology and etiology of gliomas. Acta Neuropathol. **109**, 93–108 (2005)
11. O. Warburg, On the origin of cancer cells. Science **123**, 309–314 (1956)
12. F. Pistollata, H.-L. Chen, B.R. Rood, H.-Z. Zhang, D. D'Avella, L. Denaro, M. Gardiman, G. Te Kronnie, P.H. Schwartz, E. Favaro, S. Indraccolo, G. Basso, D.M. Panchision, Hypoxia and HIF1alpha repress the differentiative effects of BMPs in high-grade glioma. Cancer Stem Cells **27**, 7–17 (2009)
13. H. Ramsahye et al., Central neurocytoma: radiological and clinico-pathological findings in 18 patients and one additional MRS case. J. Neuroradiol. (2013)
14. P.Y. Wen, S. Kesari, Malignant gliomas in adults. N. Engl. J. Med. **359**(5), 492–507 (2008)
15. J. Chen et al., A restricted cell population propagates glioblastoma growth after chemotherapy. Nature **488**(7412), 522–526 (2012)
16. FDA Approval for Temozolomide, National Cancer Institute (2010). http://www.cancer.gov/cancertopics/druginfo/fda-temozolomide
17. L. Ricci-Vitiani et al., Tumour vascularization via endothelial differentiation of glioblastoma stem-like cells. Nature **468**(7325), 824–828 (2010)
18. R. Stupp et al., Radiotherapy plus concomitant and adjuvant temozolomide for glioblastoma. N. Engl. J. Med. **352**(10), 987–996 (2005)
19. A.K. Anand et al., Survival outcome and neurotoxicity in patients of high-grade gliomas treated with conformal radiation and temozolamide. J. Cancer Res. Ther. **8**(1), 50–56 (2012)
20. S. Sahebjam et al., Bevacizumab use for recurrent high-grade glioma at McGill University Hospital. Can. J. Neurol. Sci. **40**(2), 241–246 (2013)
21. F.B. Furnari et al., Malignant astrocytic glioma: genetics, biology, and paths to treatment. Genes Dev. **21**(21), 2683–2710 (2007)
22. B.E. Stopschinski, C.P. Beier, D. Beier, Glioblastoma cancer stem cells—from concept to clinical application. Cancer Lett. **338**, 32–40 (2013)
23. B.S. Malmer et al., Genetic variation in p53 and ATM haplotypes and risk of glioma and meningioma. J. Neurooncol. **82**(3), 229–237 (2007)
24. L.E. Wang et al., Polymorphisms of DNA repair genes and risk of glioma. Cancer Res. **64**(16), 5560–5563 (2004)
25. D.A. Haas-Kogan et al., p53 function influences the effect of fractionated radiotherapy on glioblastoma tumors. Int. J. Radiat. Oncol. Biol. Phys. **43**(2), 399–403 (1999)
26. I. Zawlik et al., Common polymorphisms in the MDM2 and TP53 genes and the relationship between TP53 mutations and patient outcomes in glioblastomas. Brain Pathol. **19**(2), 188–194 (2009)
27. S.E. Yost et al., High-resolution mutational profiling suggests the genetic validity of glioblastoma patient-derived pre-clinical models. PLoS ONE **8**(2), e56185 (2013)
28. L. Bethke et al., The common D302H variant of CASP8 is associated with risk of glioma. Cancer Epidemiol. Biomark. Prev. **17**(4), 987–989 (2008)
29. A. Wigertz et al., Allergic conditions and brain tumor risk. Am. J. Epidemiol. **166**(8), 941–950 (2007)
30. K. Ueki et al., CDKN2/p16 or RB alterations occur in the majority of glioblastomas and are inversely correlated. Cancer Res. **56**(1), 150–153 (1996)

31. K. Ichimura et al., Distinct patterns of deletion on 10p and 10q suggest involvement of multiple tumor suppressor genes in the development of astrocytic gliomas of different malignancy grades. Genes Chromosomes Cancer 22(1), 9–15 (1998)

32. W. Biernat et al., Alterations of cell cycle regulatory genes in primary (de novo) and secondary glioblastomas. Acta Neuropathol. 94(4), 303–309 (1997)

33. S. Pfister et al., BRAF gene duplication constitutes a mechanism of MAPK pathway activation in low-grade astrocytomas. J. Clin. Invest. 118(5), 1739–1749 (2008)

34. D.T. Jones et al., Tandem duplication producing a novel oncogenic BRAF fusion gene defines the majority of pilocytic astrocytomas. Cancer Res. 68(21), 8673–8677 (2008)

35. A. Kaul et al., Pediatric glioma-associated KIAA1549:BRAF expression regulates neuroglial cell growth in a cell type-specific and mTOR-dependent manner. Genes Dev. 26(23), 2561–2566 (2012)

36. C. Wibom et al., EGFR gene variants are associated with specific somatic aberrations in glioma. PLoS ONE 7(12), e47929 (2012)

37. B.M. Costa et al., Impact of EGFR genetic variants on glioma risk and patient outcome. Cancer Epidemiol. Biomark. Prev. 20(12), 2610–2617 (2011)

38. S. Jeon et al., Genetic variants of AICDA/CASP14 associated with childhood brain tumor. Genet. Mol. Res. 12(AOP) (2013)

39. V.E. Clark et al., Genomic analysis of non-NF2 meningiomas reveals mutations in TRAF7, KLF4, AKT1, and SMO. Science 339(6123), 1077–1080 (2013)

40. L.M. Dyer, K.P. Schooler, L. Ai, C. Klop, J. Qiu, K.D. Robertson, D. Kevin, The transglutaminase 2 gene is aberrantly hypermethylated in glioma. J. Neurooncol. 101(3), 429 (2010)

41. A. Restrepo et al., Epigenetic regulation of glial fibrillary acidic protein by DNA methylation in human malignant gliomas. Neuro-Oncol. 13(1), 42–50 (2011)

42. C. Piperi et al., High incidence of MGMT and RARbeta promoter methylation in primary glioblastomas: association with histopathological characteristics, inflammatory mediators and clinical outcome. Mol. Med. 16(1–2), 1–9 (2010)

43. S.A. Kim et al., Promoter methylation of WNT inhibitory factor-1 and expression pattern of WNT/beta-catenin pathway in human astrocytoma: pathologic and prognostic correlations. Mod. Pathol. (2013)

44. N. Syed et al., Epigenetic status of argininosuccinate synthetase and argininosuccinate lyase modulates autophagy and cell death in glioblastoma. Cell Death Dis. 4, e458 (2013)

45. P. Vaitkiene et al., GATA4 and DcR1 methylation in glioblastomas. Diagn. Pathol. 8(1), 7 (2013)

46. M. Martini et al., Epigenetic silencing of Id4 identifies a glioblastoma subgroup with a better prognosis as a consequence of an inhibition of angiogenesis. Cancer 119(5), 1004–1012 (2013)

47. A.P. Chou et al., Identification of retinol binding protein 1 promoter hypermethylation in isocitrate dehydrogenase 1 and 2 mutant gliomas. J. Natl. Cancer Inst. 104(19), 1458–1469 (2012)

48. A. von dem Knesebeck et al., RANK (TNFRSF11A) is epigenetically inactivated and induces apoptosis in gliomas. Neoplasia 14(6), 526–534 (2012)

49. D. Skiriute et al., MGMT, GATA6, CD81, DR4, and CASP8 gene promoter methylation in glioblastoma. BMC Cancer 12, 218 (2012)

50. A. Waha et al., Frequent epigenetic inactivation of the chaperone SGNE1/7B2 in human gliomas. Int. J. Cancer 131(3), 612–622 (2012)

51. M. Kadowaki et al., DNA methylation-mediated silencing of nonsteroidal anti-inflammatory drug-activated gene (NAG-1/GDF15) in glioma cell lines. Int. J. Cancer 130(2), 267–277 (2012)

52. M. Wetzel et al., Effect of trichostatin A, a histone deacetylase inhibitor, on glioma proliferation in vitro by inducing cell cycle arrest and apoptosis. J. Neurosurg. 103(6 Suppl), 549–556 (2005)

53. A. Bangert et al., Chemosensitization of glioblastoma cells by the histone deacetylase inhibitor MS275. Anticancer Drugs 22(6), 494–499 (2011)

54. A. Bangert et al., Histone deacetylase inhibitors sensitize glioblastoma cells to TRAIL-induced apoptosis by c-myc-mediated downregulation of cFLIP. Oncogene **31**(44), 4677–4688 (2012)
55. S. Hacker et al., Histone deacetylase inhibitors prime medulloblastoma cells for chemotherapy-induced apoptosis by enhancing p53-dependent Bax activation. Oncogene **30**(19), 2275–2281 (2011)
56. R.A. Gatenby, R.J. Gillies, Why do cancers have high aerobic glycolysis? Nat. Rev. Cancer **4**(11), 891–899 (2004)
57. R.J. DeBerardinis et al., The biology of cancer: metabolic reprogramming fuels cell growth and proliferation. Cell Metabolism **7**(1), 11–20 (2008)
58. D. Hanahan, R.A. Weinberg, The hallmarks of cancer. Cell **100**(1), 57–70 (2000)
59. D. Hanahan, R.A. Weinberg, Hallmarks of cancer: the next generation. Cell **144**(5), 646–674 (2011)
60. A. Ramanathan, C. Wang, S.L. Schreiber, Perturbational profiling of a cell-line model of tumorigenesis by using metabolic measurements. Proc. Natl. Acad. Sci. **102**(17), 5992–5997 (2005)
61. I. Marin-Valencia et al., Analysis of tumor metabolism reveals mitochondrial glucose oxidation in genetically diverse human glioblastomas in the mouse brain in vivo. Cell Metabolism **15**(6), 827–837 (2012)
62. T.N. Seyfried, L.M. Shelton, P. Mukherjee, Does the existing standard of care increase glioblastoma energy metabolism? Lancet Oncol. **11**, 811–813 (2010)
63. T.N. Seyfried, L.M. Shelton, Cancer as a metabolic disease. Nutr. Metab. **7**, 7 (2010)
64. T.N. Seyfried, M.A. Kiebish, J. Marsh, L.M. Shelton, L.C. Huysentruyt, P. Mukherjee, Metabolic management of brain cancer. Biochim. Biophys. Acta **1807**, 577–594 (2011)
65. E.A. Maher et al., Metabolism of [U-13C] glucose in human brain tumors in vivo. NMR Biomed. **25**(11), 1234–1244 (2012)
66. E.A. Maher, F.B. Furnari, R.M. Bachoo, D.H. Rowitch, D.N. Louis, W.K. Cavenee, R.A. DePinho, Malignant glioma: genetics and biology of a grave matter. Genes Dev. **15**, 1311–1333 (2001)
67. K.M. Egan et al., Cancer susceptibility variants and the risk of adult glioma in a US case-control study. J. Neurooncol. **104**(2), 535–542 (2011)
68. D. Hanahan, R.A. Weinberg, In search of cancer's common ground: a next-generation view. ScienceDaily (2011). http://www.sciencedaily.com/releases/2011/03/110303132300.htm
69. A. Carter, Curry compound fights cancer in the clinic. J. Natl. Cancer Inst. **100**, 616–617 (2008)
70. J. Ravindran, S. Prasad, B.B. Aggarwal, Curcumin and cancer cells: how many ways can curry kill tumor cells selectively? AAPS J. **11**, 495–510 (2009)
71. W.-Y. Huang, Y.-Z. Cai, Y. Zhang, Natural phenolic compounds from medicinal herbs and dietary plants: potential use for cancer prevention. Nutr. Cancer **62**(1), 1–20 (2009)
72. S. Purkayastha et al., Curcumin blocks brain tumor formation. Brain Res. (2009)
73. S. Shishodia, G. Sethi, B.B. Aggarwal, Curcumin: getting back to the roots. Ann. N.Y. Acad. Sci. **1056**(1), 206–217 (2005)
74. J. Weissenberger et al., Dietary curcumin attenuates glioma growth in a syngeneic mouse model by inhibition of the JAK1,2/STAT3 signaling pathway. Clin. Cancer Res. **16**(23), 5781–5795 (2010)
75. W. Zhuang et al., Curcumin promotes differentiation of glioma-initiating cells by inducing autophagy. Cancer Sci. **103**(4), 684–690 (2012)
76. C. Senft et al., The nontoxic natural compound Curcumin exerts anti-proliferative, anti-migratory, and anti-invasive properties against malignant gliomas. BMC Cancer **10**, 491 (2010)
77. T.Y. Huang et al., Curcuminoids suppress the growth and induce apoptosis through caspase-3-dependent pathways in glioblastoma multiforme (GBM) 8401 cells. J. Agric. Food Chem. **58**(19), 10639–10645 (2010)

78. M.C. Perry et al., Curcumin inhibits tumor growth and angiogenesis in glioblastoma xenografts. Mol. Nutr. Food Res. **54**(8), 1192–1201 (2010)
79. S. Karmakar, N.L. Banik, S.K. Ray, Curcumin suppressed anti-apoptotic signals and activated cysteine proteases for apoptosis in human malignant glioblastoma U87MG cells. Neurochem. Res. **32**(12), 2103–2113 (2007)
80. A.K. Khaw et al., Curcumin inhibits telomerase and induces telomere shortening and apoptosis in brain tumour cells. J. Cell Biochem. (2012)
81. A. Goel, S. Jhurani, B.B. Aggarwal, Multi-targeted therapy by curcumin: how spicy is it? Mol. Nutr. Food Res. **52**, 1010–1030 (2008)
82. P. Langone, G.M. Curcio, K. Kashfi, S. Dolai, K. Raja, P. Banerjee, Drug targeting to eliminate breast and brain tumors, in *Joint AACR and ACS Meeting—Chemistry in Cancer Research: The Biological Chemistry of Inflammation as a Cause of Cancer, San Diego, CA* (2011)
83. S.J. Lee et al., Curcumin-induced HDAC inhibition and attenuation of medulloblastoma growth in vitro and in vivo. BMC Cancer **11**, 144 (2011)
84. S.K. Kang, S.H. Cha, H.G. Jeon, Curcumin-induced histone hypoacetylation enhances caspase-3-dependent glioma cell death and neurogenesis of neural progenitor cells. Stem Cells Dev. **15**(2), 165–174 (2006)
85. J. Kang et al., Curcumin-induced histone hypoacetylation: the role of reactive oxygen species. Biochem. Pharmacol. **69**(8), 1205–1213 (2005)
86. K. Balasubramanyam et al., Curcumin, a novel p300/CREB-binding protein-specific inhibitor of acetyltransferase, represses the acetylation of histone/nonhistone proteins and histone acetyltransferase-dependent chromatin transcription. J. Biol. Chem. **279**(49), 51163–51171 (2004)
87. J. Fang, J. Lu, A. Holmgren, Thioredoxin reductase is irreversibly modified by curcumin: a novel molecular mechanism for its anticancer activity. J. Biol. Chem. **280**, 25284–25290 (2005)
88. C. Syng-Ai, A.L. Kumari, A. Khar, Effect of curcumin on normal and tumor cells: role of glutathione and bcl-2. Mol. Cancer Ther. **3**, 1101–1108 (2004)
89. B.B. Aggarwal, B. Sung, Pharmacological basis for the role of curcumin in chronic diseases: an age-old spice with modern targets. Trends Pharmacol. Sci. **30**, 85–94 (2009)
90. S. Manju, K. Sreenivasan, Enhanced drug loading on magnetic nanoparticles by layer-by-layer assembly using drug conjugates: blood compatibility evaluation and targeted drug delivery in cancer cells. Langmuir **27**(23), 14489–14496 (2011)
91. J. Shao et al., Curcumin delivery by methoxy polyethylene glycol-poly(caprolactone) nanoparticles inhibits the growth of C6 glioma cells. Acta Biochim. Biophys. Sin. (Shanghai) **43**(4), 267–274 (2011)
92. P. Langone, P.R. Debata, S. Dolai, G.M. Curcio, J.D. Inigo, K. Raja, P. Banerjee, Coupling to a cancer cell-specific antibody potentiates tumoricidal properties of curcumin. Int. J. Cancer **131**, E569–E578 (2012)
93. Drug Record: Temozolomide, in *Clinical and Research Information on Drug-Induced Liver Injury* (2012). NIDDK, http://livertox.nih.gov/Temozolomide.htm#overview
94. T. Takano, J.H. Lin, G. Arcuino, Q. Gao, J. Yang, M. Nedergaard, Glutamata release promotes growth of malignant gliomas. Nat. Med. **7**, 1010–1015 (2001)
95. R. Grondin, Z. Zhang, Y. Ai, D.M. Gash, G.A. Gerhardt, Intracranial delivery of proteins and peptides as a therapy for neurodegenerative diseases. Prog. Drug Res. **61**, 101–123 (2003)
96. D. Grady, A direct hit of drugs to treat brain cancer (2010). Available from http://www.nytimes.com/2010/11/09/health/09avastin.html

To Those Who Touched My Life

Don't stand over there and cry
There is not, where you can find me
Though I did die, I am here
I am the rustle in the branches of the redwood trees
I am the sunbeam on the humming bird's wings
I am the waterfall in the mountain
I am the silence before sunset over the ocean
I am the stars that tell the story of the universe
I am here and will always be
As long as you remember.

Hisako Matsubara
Mino's mother

F. Freund, S. Langhoff (eds.), *Universe of Scales: From Nanotechnology to Cosmology*,
Springer Proceedings in Physics 150, DOI 10.1007/978-3-319-02207-9,
© Springer International Publishing Switzerland 2014

CPSIA information can be obtained at www.ICGtesting.com
Printed in the USA
BVOW11*0005250714

360450BV00001B/1/P